SPACE
THE True Frontier!

Or When Is Grace Not Enough?

To my loving wife.
May 25/2018
Love Eric

ERIC RASMUSSEN

ISBN 978-1-64258-926-9 (paperback)
ISBN 978-1-64258-927-6 (digital)

Copyright © 2018 by Eric Rasmussen

All rights reserved. No part of this publication may be reproduced, distributed, or transmitted in any form or by any means, including photocopying, recording, or other electronic or mechanical methods without the prior written permission of the publisher. For permission requests, solicit the publisher via the address below.

Christian Faith Publishing, Inc.
832 Park Avenue
Meadville, PA 16335
www.christianfaithpublishing.com

Printed in the United States of America

To my friend George, who said, "Write it down."

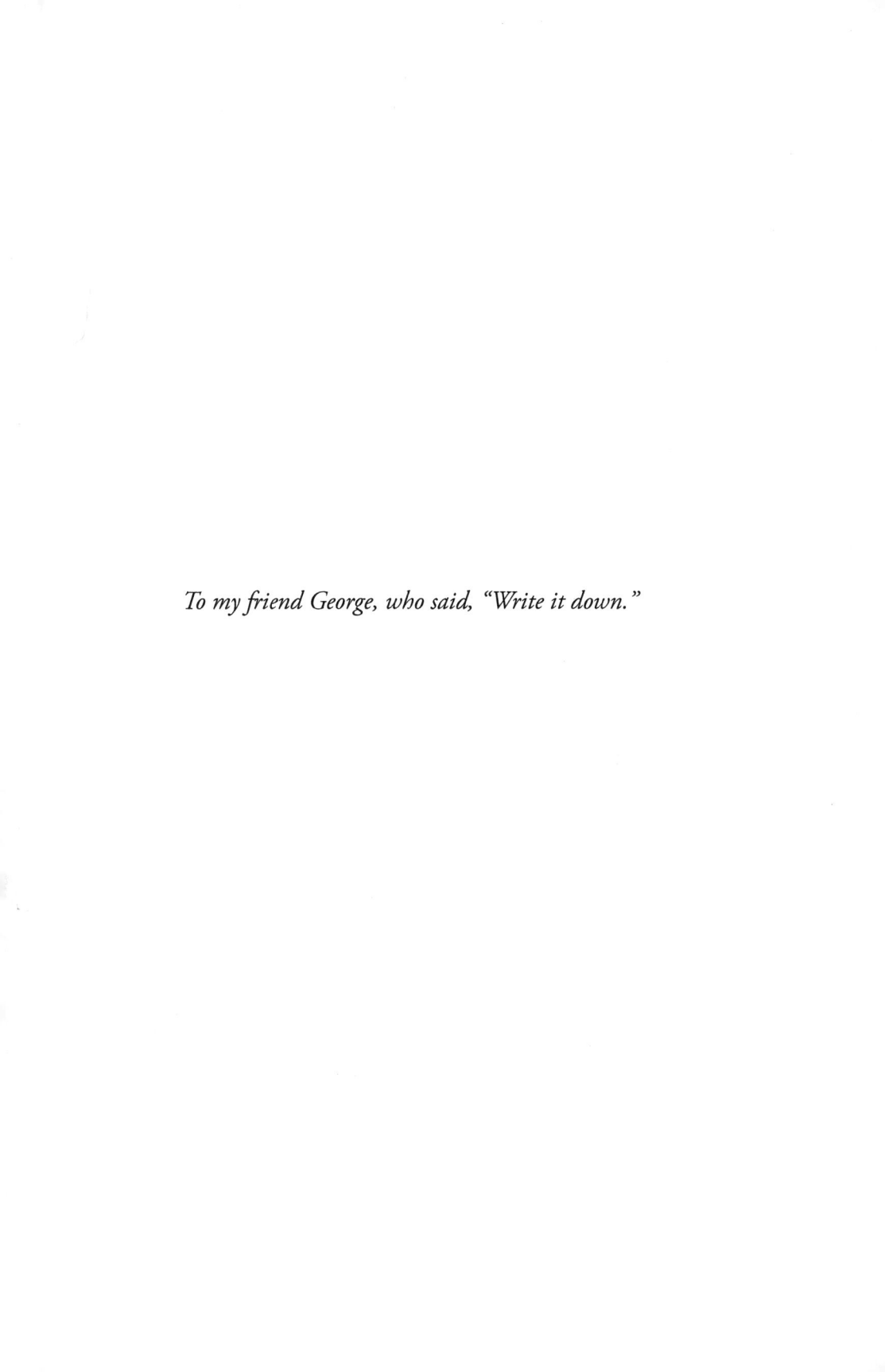

The little book that wrote itself.
It was the hardest thing I have ever done.
I could only have done it because I am certain in my heart God wanted me to.

Eric Rasmussen
November 1, 2017

CONTENTS

Introduction 9

Chapter 1: The Parable of the Sower 79

Chapter 2: Not Three 89

Chapter 3: Hiccups Along the Road to the Universe ... 150

Chapter 4: The Not Forgotten 164

Chapter 5: Knock, Knock! 185

Chapter 6: My Credentials 211

Epilogue 221

Credits 225

Appendices 227

INTRODUCTION

Space, the true frontier!

How can something so ubiquitous as space not have everything to do with us?

I say it is our inheritance, our garden that we will dress and beautify, the future venue of the family of God.

God exists in the glory of space today, and for us it is coming.

> I consider that our present sufferings are not worth comparing with the glory that will be revealed in us. The creation waits in eager expectation for the sons of God to be revealed. For the creation was subjected to frustration, not by its own choice, but by the will of the one who subjected it, in hope that the creation itself will *be liberated from its bondage to decay and brought into the glorious freedom of the children of God.* (Romans 8:18–21)

Psalms 19:1 (NIV) The heavens declare the glory of God; the skies proclaim the work of his hands.

Romans 1:18 (NIV) [18] The wrath of God is being revealed from heaven against all the godlessness and wickedness of men who suppress the truth by their wickedness, [19] since what may be known about God is plain to them, because God has made it plain to them. [20] For since the creation of the world God's invisible qualities—his eternal power and divine nature—have been clearly seen, being understood from what has been made, so that men are without excuse.

Below are some pictures, of that God. At least they reveal what He is like.

Lake Louise, Moraine Lake, the Pillars of Creation, and the Ultra Deep Field, show His glory.

They did not come from nothing, …slowly.

It took a great genius beyond understanding.

Philippians 4:4 (NIV)

[4] Rejoice in the Lord always. I will say it again: Rejoice! [5] Let your gentleness be evident to all. The Lord is near. [6] Do not be anxious about anything, but in everything, by prayer and petition, with thanksgiving, present your requests to God. [7] And the peace of God, which transcends all understanding, will guard your hearts and your minds in Christ Jesus.

[8] Finally, brothers, whatever is true, whatever is noble, whatever is right, whatever is pure, whatever is lovely, whatever is admirable—if anything is excellent or praiseworthy—think about such things. [9] Whatever you have learned or received or heard from me, or seen in me—put it into practice. And the God of peace will be with you.

God is making us His family, we shall be like Him. Like this?

Figure 1 Lake Louise Alberta Canada
Source: Internet 4k pics

SPACE THE TRUE FRONTIER!

Figure 2 Moraine Lake Alberta Canada
Source: Internet 4k pics

If I read that correctly, the glorious-divine nature is revealed, in the magnificence of the created earth, and stars above, what we call, 'space'.

Figure 3 Pillars of Creation-showing birth of stars
Source: Internet Hubble 4k pics

Figure 4 Space
Source: Internet Hubble 4k Ultra Deep Field Pics

Psalms 8:3 (NIV) [3] When I consider your heavens,
the work of your fingers,
the moon and the stars,
which you have set in place,
[4] what is man that you are mindful of him,
the son of man that you care for him?
[5] You made him a little lower than the heavenly beings
and crowned him with glory and honor.
The heavens declare the glory of God, and He has
promised to share that glory with us!
John 17: [5] Now, Father, glorify Me in Your presence
with that glory I had with You
before the world existed.
John 17: [20] I pray not only for these,
but also for those who believe in Me
through their message.
[21] May they all be one,
as You, Father, are in Me and I am in You.

SPACE THE TRUE FRONTIER!

May they also be one in Us,
so the world may believe You sent Me.
[22] I have given them the glory You have given Me.
May they be one as We are one.

The Bible reveals to us that God is sharing His existence by making us adopted sons in his family, and if we are to become part of His oneness, it begs the questions **what and where is God** and, even more wondrous, **when is God**?

(If God sees from all frames of reference instantaneously, then He is "everywhen" or "anywhen" and outside of time that He created?)

I believe, and will show later, that biblically, Jesus is the Creator of the **space-time continuum** we find ourselves existing in.

I also believe that science is giving us fantastic insight into how God did it. Right down to the very limits of the strange world of quantum mechanics.

Researchers capture image of hydrogen atoms electron orbital for first time.

Figure 5 Quantum Picture of a Hydrogen Atom
Source: https://www.geek.com/news/researchers-capture-image-of-hydrogen-atoms-electron-orbital-for-first-time-1556448/

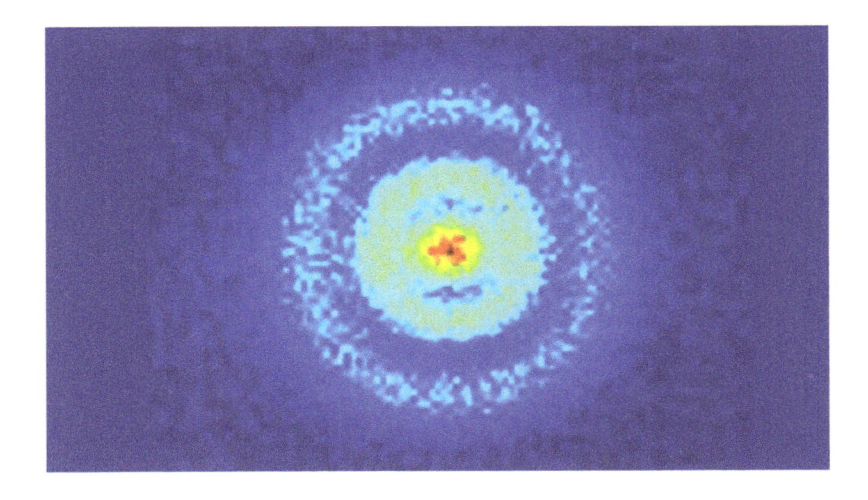

Astonishingly, I think they **are looking increasingly closely at God Himself**. And perhaps, the only difference between science and Christianity regarding these new truths is the perceived source of what they are finding. The observations, and the resulting standard-model reality, are true for both science and Christians.

Could God be the instantaneous "entanglement" that happens between separated subatomic particles,—I won't say simultaneously—but everywhen?

And is it God doing this, no matter how far apart they might be, because God is everywhere as everything?[1]

I suspect the Large Hadron Collider in Bern Switzerland is indeed seeing the source of matter in the universe, with science understanding it to be the unintelligent Higgs boson field and myself betting it is the ultrapowerful, all-intelligent God they are examining so closely. The new reality that everyone must concede is that there are photons of light and that they pass through double-slit experiments occurring all over the world these days. Something made these photons, and something gives these photons the apparent insight to choose how to act. When there are two slits, they show wave interference; if only one slit, they act as a particle. By the way, the Bible says, "God is Light." Interesting.[2]

If God is everywhere and everything and everywhen, then He would know the number of slits, and what pattern to make on the other side, and even if someone dared to take a peek?

One of the strangest things about the double-slit experiment is that if you look to see when the photons pass through (if you peek), you change the outcome! The ultimate question is, what told the photon you were watching? Nothing is faster than light, so what caught up to it, and said, 'hey someone is watching, so act like a particle this time.

Science has a prejudice—that our existence cannot come from intelligence but must (atheistically) be founded with unintelligence, assisted by unimaginable—heaps of chance.

And I must admit, I too have a prejudice. I have a faith that our existence requires an intelligent, all-powerful creator. This foundation came to me through miracles in my life. It is the only way I would have believed in a creator and overcome my early indoctrination into the god of science—evolution.

God had to open my eyes.

These were personal events, so I don't expect you to understand this, but for me, the Bible is now a very special book that is from God. And I am thankful we truly are not alone. God is watching, and God will lovingly save us (each in our time), bringing all mankind into His family and His glorious universe (fully accessible space-time).

But also, using scientific fundamentals, I base this faith to a large degree on the observational evidence that entropy (winding downness) can only be reversed by intelligent energetic involvement. So the biggest windup of all, the big bang of it, should require no less intelligence.

Okay, all that aside, it doesn't mean what atheistic science discovers isn't fantastic and useful information to everyone, no matter how things got here.

I say, "**Thank you, science**, for giving us such a high-pixel clarity of what God is." I say to Christians, "Pay attention. This picture being painted, these standard model particles exist, just as science is finding. So, Christians, what they are revealing must be how God did it."

I wonder if science and religion aren't like East and West. If you hold out a string and hold the east and west ends, it is true that east is as far away from west as possible. However; when you wrap that string around a globe, the extremities come together and are joined at the same point. Perhaps science and religion are like that, and have all along been heading to the same truth of existence. Heading to the same point, just coming from the opposite direction?

As I alluded, I feel, or suspect, **space**—all that physically exists—is a kind of frozen state of His Spirit? Something that **is Him**, His essence Spirit in a different solid state of being? So for me, this ever-focusing picture is revealing Gods physical manifestation of Himself.

Like some scientists believe, the universe is not only described by math but, in fact, is math![3]

I think they are saying something close to what I feel. Only I am saying, **everything is the mathematician (God)** and not just the math. Literally.

> *"Am I only a God nearby,"* declares the LORD, *"and not a God far away? Can anyone hide in secret places so that I cannot see him?"* declares the LORD. ***"Do not I fill heaven and earth?"* declares the LORD.** *(Jeremiah 23:23–24, NIV)*

> ***Where can I go from your Spirit?*** *Where can I flee from your presence?* ***If I go up to the heavens, you are there;*** *if I make my bed in the depths, you are there.* *(Psalms 139:7–8, NIV)*

That's a lot to take in, and you don't have to believe this, for the results of this book to be at one with what God is doing.

My pet idea that everything is made of God particles, speculative as it is, keeps growing on me, becoming more plausible as life goes on. So I hope you might consider it, simply because I like to share what I think God is telling me. It brings me joy and keeps coming to focus in my heart, and I don't believe it harmful to see it that way.

Also, it nicely explains some of the problems physicists are confronted with over the last century (entanglement and faster-than-light double-slit photon communication). If matter is God, then God knows because He is everything, everywhere, anywhen.

The scriptures say that God is Love, God is Spirit, God is Light. If He is all that and from Him came the "dirt," then would He not be all that is spiritual and all that is physical at the same space-time?[4]

Like ice is a different state of water, could matter simply be a different state of spirit? I think so.

SPACE THE TRUE FRONTIER!

> Water appears in nature in all **three** common states of matter (solid, liquid, and gas) and may take many different forms on Earth: water vapor and clouds in the sky, seawater in the oceans, icebergs in the polar oceans, glaciers in the mountains, fresh and salt water lakes, rivers, and aquifers in the ground.
>
> Water - Wikipedia
> en.wikipedia.org/wiki/Water

And Jesus says He is our water! He passes it on to us, and we pass it on to others![5]

> *On the last and greatest day of the Feast, Jesus stood and said in a loud voice, "If anyone is thirsty, let him come to me and drink. **Whoever believes** in me, as the Scripture has said, **streams of living water will flow from within him.**" (John 7:37–38, NIV)*

If there was only God before all else, then wouldn't everything He makes be part of Him including the physical?[6] Jesus was both man and God, physical and spiritual.

And just like we are **not** contained within anything we might make, at the same time God is not restrained, or contained, within the limits of His physical creation (the universe and time).

What I am pointing out is that some of the statements God made in the Bible, have become plausible in a literal sense, thanks to the understanding of the universe revealed through science (relativity and quantum mechanics) in the last century. *Second Peter 3:8 (YLT): "And this one thing let not be unobserved by you, beloved, that one day with the Lord is as a thousand years, and a thousand years as one day."*[7]

And if this idea that our omnipresent God is everything physical (as frozen spirit), as well as everything spiritual, then God would be **everywhere, everything**, and most interestingly, **everywhen**![8]

Now if we become part of that, then space and time will be ours too! And I want to show that is our incredible promise!

Eventually, we shall be like Him, existing eternally outside of time, making space available to us as a new and true spiritual garden of Eden. And God promises these new heavens will be **liberated from bondage and decay (they will then be eternal**—entropy will no longer happen).[9]

He also says,

> "In the beginning, O Lord, you laid the foundations of the earth,
> *and the heavens are the work of your hands.*
> They will perish, but you remain;
> *they will all wear out like a garment.*
> You will roll them up like a robe;
> like a garment they will be changed.
> But you remain the same, and your years will never end." (Hebrews 1:10–12)

I propose space-time is the true frontier prepared for us before time. Our true place is outer space.
So far, we are physically part of that, with a down payment only of the fullness of His likeness, later born again into the full glory of His spiritual existence. **(And for science that is searching so hard, if my ideas are sound, then indeed, God is the Higgs boson field and every particle is a God particle).** ==My hope with this book is to show that God is sharing His existence, which includes space and time and a future new creation that will be eternal!== And I know, that it will at first be considered ridiculous then highly resisted, but I think it will eventually become obvious.

God's plan has always been considered foolishness to the world and cannot be seen correctly (in its glorious reality) without God first removing His **protective blindness.** *John 9:41: "Jesus said, 'If you were blind, you would not be guilty of sin; but now that you claim you can see, your guilt remains.'"* Only the Father knows the best time to tell us and the best time to show us (to give us hearing and sight). What an awesome God.

For the rest of my introduction, I have prepared a "*Readers Digest* version" of the entire story and back it up later in the chapters using more detail (repetition is a great teacher).

I am doing this because I want to amplify how God has always intended that "our true place is outer space." So I am sharing scriptures that expand on the idea that space is both the home of God and is excitingly our future glorious venue. These things could not be as fully understood over the course of much of Christian history until the understanding and increased knowledge of our present times. And fantastically, I hope we can realize that these ancient writers are speaking about truths that has taken science millennia to agree with.

The Word of God talked about our vast universe long before there was science to confirm it!

I want to make that point stand out! These ancient words are not a typical understanding to be expected from the writings of near cave men. Like the Bible says, they are inspired revelation from the creator that purposed it!

People, on their own, without God's input, have understandably come up with colorful but ignorant ideas, like "the night sky is light peeking through holes in a canopy of some kind." How could man know any better back then, but God knew, didn't He?[10] These ancient God-inspired writings tell us about a limitless glorious universe that has a God who stands above their circuits, don't they? The words of the Bible regarding the universe are still at one with the latest discoveries of science. Now that is something to consider from a book that is thousands of years old.

Then I will move on to chapter 1, where God explains how and why **this is a hidden mystery**. And God explains that we are blind to it until our eyes are opened. True for unbelievers and Christians to varying degrees.

And I think it is time to present another question. And that question is, **why did God bother with a physical universe if our ultimate destiny is spiritual** existence in His family? If God wanted us in heaven (the glorious universe), why didn't He just put us there?

Well, He did something close to that with another being, one much greater than we now are— the original archangel light bringer that rebelled and became Satan. He no longer has the role of light-bringer but is now darkness and anti-God. He is also spiritual and will exist forever in his miserably unhappy state, unfortunately for him.

I will show from the Bible this great spiritual rebellion against God's plan, where the great archangel lost his place in the universe, becoming Satan. He attempted to destroy everything but was overpowered and subsequently thrown down to earth. This played into the plan of God because each of us must choose. Do we listen to this prince of the power of the air or to God? For a time, we all fail, but although physical people may have been deceived and had forgotten God (and unknowingly believed Satan), God has not forgotten us, the people of His promise (His family). God will bring us back to Him.

This rebellion was no surprise to God; it was known from before time. For it is stated that since before creation, it was understood to be necessary that the Lamb had to be slain for salvation. I suspect this rebellion is the reason that when God moved upon the face of the waters in Genesis, it had become a scene of destruction, requiring a re-creation of the earth (for the coming domain of mankind).

Becoming part of the life of God seems to require choice. And a wrong choice can be erased as long as we are only physical.

That is my guess for our first existence being only a physical reality. Because we are now only physical beings, if in the end we hate God and what he is doing, then we can be as if we never existed in the first place. That would mean, we have no part in the "tree of life" and glorious eternity.

> *But it is the spirit in a man, the breath of the Almighty, that gives him understanding. (Job 32:8, NIV)*

> *The Spirit of God has made me; the breath of the Almighty gives me life. (Job 33:4, NIV)*

> *And the LORD God formed man of the dust of the ground, and breathed into his nostrils the breath of life; and man became a living soul. (Genesis 2:7, KJV)*

Just one breath of God's Personal Spirit in us brought the dust we are to life as a physical soul (man).

> *The soul that sinneth, it shall die. The son shall not bear the iniquity of the father, neither shall the father bear the iniquity of the son: the righteousness of the righteous shall be upon him, and the wickedness of the wicked shall be upon him. (Ezekiel 18:20, KJV)*

> *Have I any pleasure at all that the wicked should die? saith the Lord GOD: and not that he should return from his ways, and live? (Ezekiel 18:24, KJV)*

> *For the wages of sin is death, but the gift of God is eternal life in Christ Jesus our Lord. (Romans 6:23, HCSB)*

Unlike Satan (who some might choose to follow and worship), we could then be mercifully erased because we are still only physical. Could the second death talked about in Revelation be the erasure?

> *Blessed and holy are those who have part in the first resurrection. **The second death** has no power over them, but they will be priests of God and of Christ and will reign with him for a thousand years. (Revelation 20:6, NIV)*

> *Do not be afraid of those who kill the body but cannot kill the soul. Rather, **be afraid of the One who can destroy both soul and body in hell.** (Matthew 10:28, NIV)*[11]

The first death that we all experience is physical death. The Bible shows it is only temporary and is called sleeping. Christ resurrected several people in His day and always explained He was just waking them from sleep (early). When Christ died, there was an earthquake, the tombs were opened, and many righteous people were brought back to a physical life. And they all later died again, because they were still only physical.

Christ is the only one to have been resurrected to a spirit being so far. Man, if approved by God, will become spirit at the return of Christ. Those in the grave who are still sleeping at that time and subsequent to that and those who are alive on earth at that time will be resurrected at the last trump (Mark 5:35–43; 1 Thessalonians 4:13–18).

But the second death is not the desire of God, with his tremendous grace. Thank God for Christ, who is our graceful salvation. And remember God's will for us is sure. What God wants, God will get. So let us focus on that love prevailing and <u>choose</u> daily to follow the Spirit of God (Christ and the Father in us). May we always say, "In the name of Jesus and by His power, get behind us, Satan."

As a Berean (Acts 17:11), I will share how I see the God head, and future kingdom and family of God. I believe the deceiver has fooled even Christians into becoming deceived colleagues, rewriting Gods plan, making it into Satan's plan.

Our Savior Christ was tempted with this counterfeit but not fooled into becoming part of it, and Christ thereby defeated Satan, as must we. Satan confronted Christ as if he was the Father God himself, able to put Christ into the kingdom without any need for Christ to suffer.

Christ did not listen and thereby did not worship Satan.

<u>The Creator of everything humbled Himself to almost nothing</u> to make Gods plan succeed. The only way man could defeat Satan is for God to become man! More on that later.

I humbly, yet boldly, state that deception hiding our part in this plan must and will be revealed by the authority of Christ. May God bless us all with our successful inheritance, through Christ, into the likeness and into the very family of God.

Now more on what we inherit, space-time: It has been in our Bible for thousands of years, just as the Hubble Space Telescope is now observing.

Keep this question in the back of your mind: where is God, **who will eventually be there with Him and when**?

> *Is not God high in heaven? And see the summit of the stars, That they are high. And thou hast said, "What--hath God known? Through thickness doth He judge? Thick clouds are a secret place to Him, And He doth not see;"* **And the circle of the heavens He walketh habitually**. *(Job 22:12–14, YLT)*

The picture below is what I would call an amazing understanding for these ancient writers; however, this may not be what the Hebrew is directly referring to.[12] Therefore, this passage may or may not be showing an early revelation of our spherical, and orbital heavens. Either way, sphere, circle, or vault, the scripture is accurate and does not go against modern science. It would have been an understandable error for the writers of that time to have a flat-earth view. It would also be understandable for- people of the time to talk about candles of light shining through the holes in the canvas of the tent above, rather than to show the understanding of God in a universe of stars above the earth.

Figure 6. Images of the circle (sphere) of the earth (Bing.com/images)

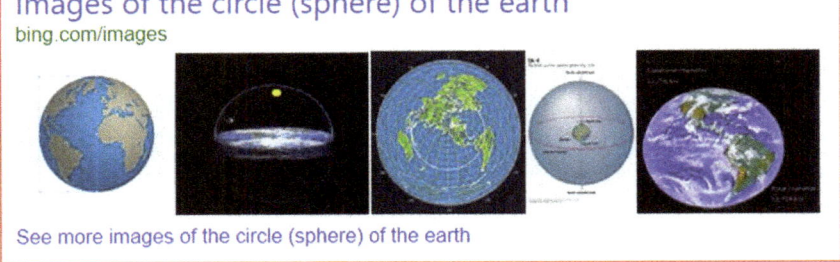

And for clarity, I am showing the Message and the Holman Christian Standard Bible translations:

> *You agree, don't you, that God is in charge?*
> *He runs the universe—just look at the stars!*
> *Yet you dare raise questions: "What does God know?*
> *From that distance and darkness, how can he judge?*
> *He roams the heavens wrapped in clouds,*
> *so how can he see us?" (Job 22:12–14, MSG)*

> *Isn't God as high as the heavens?*
> *And look at the highest stars—how lofty they are!*
> *Yet you say: "What does God know?*
> *Can He judge through thick darkness?*
> *Clouds veil Him so that He cannot see,*
> **as He walks on the circle of the sky."**
> *(Job 22:12–14, HCSB)*

Our Comforter speaks to us of a great salvation, and it will include His glorious universe that He circles about. Like John the Baptist prepared the way, by quoting these ancient words of the promise, I hope to show this is your promise and all mankind's.

Our Comforting shepherd will share His wonderful glory with us (Israel).[13] Amazing that the knowledge of the glorious universe has been in the word of God for millennia! Science take note.

I find I must also say, Christianity take note, because from what I see, the scriptures are not looked upon with the same respect as God intended. The holy scriptures (Isaiah 40) talked about in the quote below are from the Old Testament because there was no New Testament back then (they were letters at that time). Also, if God said it, I think He means it forever. God does not fail and start over! The promises He makes are solid truth, and you can trust Him when He says it! He has not abandoned us to life on our own without direction.

That is why He gave us His holy word (the Bible). And that Bible **speaks of promises that transcend both covenants**, and **the means to obtaining these promises are similar but now spiritual**.

The first covenant was nullified because it depended partly on man (and mankind failed). The new covenant cannot fail because it depends on Christ alone (and so the same promises are now spiritually certain).

The new covenant—with its New Jerusalem, new Pentecost, new temple, and new heavens—is assured because similarly our Passover Lamb of God (Jesus) was sacrificed to enable our atonement with the Father.

And the harvests, expounded in the parable of the sower, grows us into an ever-increasing spiritual harvest (including the wave-sheaf Christ, the firstfruits, and a greater multitude harvested later), that is, the family of God!

Jesus rode on two donkeys that year. For me, it was a picture of Christ making the transition into the new covenant from the old. Christ was born under the law of Moses as a Jew, but He died as the New Testament fulfillment, as Messiah of all Israel (new and old). There will be no different Messiah coming. Christ, He is risen.

I propose the idea that He rode on an old donkey to get to Jerusalem (representing the old covenant part of Christ's life), and He rode on a new donkey, a colt, into the east gate of the temple in Jerusalem in AD 31, as He and the world entered into the new covenant part of Christ's and our lives. I show elsewhere that Christ, our Passover, probably died at the exact time the priests were killing the many physical lambs that year at the temple in Jerusalem, which was a perfect and godly masterful fulfillment of God's word that we have all overlooked.[14]

The Triumphal Entry

*As they approached Jerusalem and came to Bethphage on the Mount of Olives, Jesus sent two disciples, saying to them, "Go to the village ahead of you, and at once you will find **a donkey** tied there, **with her colt by her**. Untie them and **bring them to me**. anyone says*

*anything to you, tell him that **the Lord needs them**, and he will send them right away."*
This took place to fulfill what was spoken through the prophet:

> *"Say to the Daughter of Zion,*
> *'See, your king comes to you,*
> *gentle and riding on a donkey,*
> *on a colt, the foal of a donkey.'"*

The disciples went and did as Jesus had instructed them. They brought the donkey and the colt, placed their cloaks on them, and Jesus sat on them. A very large crowd spread their cloaks on the road, while others cut branches from the trees and spread them on the road. The crowds that went ahead of him and those that followed shouted,

> *"Hosanna to the Son of David!"*
> *"Blessed is he who comes in the name of the Lord!"*
> *"Hosanna in the highest!"*

When Jesus entered Jerusalem, the whole city was stirred and asked, "Who is this?"
The crowds answered, "This is Jesus, the prophet from Nazareth in Galilee." (Matthew 21:1–11, NIV, cf Mark 11:1–10; Luke 19:29–38, 21:4–9 pp; John 12:12–15)

Jesus kept the Passover of the Old Testament that year by becoming its fulfillment, as He died enacting His role as the actual Passover Lamb (singular, wave-sheaf like, firstfruits) Himself of the new covenant.

Shortly thereafter, He becomes the first Man resurrected to spiritual life, followed later with the start of the church at Pentecost (counting fifty days from the morrow after the weekly Sabbath, during the week of the Passover, or Days of Unleavened Bread) by us. Yes us, those adopted sons, the firstfruits of the transition church, out from the Old Testament congregation. This occurred to the best of my knowledge on Sunday, Pentecost AD 31.

This New Testament church is temporarily still physical, but we have the baptism and the laying on of hands, which initiates our conception, eventually leading us to our new birth into the spiritual family of God. Remember the ubiquitous foreshadowing (physical then spiritual) that comes from our first birth into our physical families today. It is just like the holy days, foreshadowed the holy realities to come in the new covenant.

We have only the down payment of that spirit oneness now, but the completion of our being born again shall be like a big bang at the trumpet call to our change, as we are fully born again into the promised spiritual oneness with Christ and the Father.

The transition from the Old to the New Testament was a hand-off touching and bringing together both covenants—a masterful, praiseworthy springboard-like stroke of absolute genius (that is hard for us to see). The transition is pictured in Matthew 17, and I think it fulfills the last scripture from Matthew 16. *"I tell you the truth, some who are standing here will not taste death before they see the Son of Man coming in his kingdom" (Matthew 16:28, NIV).*

> *After six days Jesus took with him Peter, James and John the brother of James, and led them up a high mountain by themselves. There he was transfigured before them. His face shone like the sun, and his clothes became as white as the light. Just then there appeared before them **Moses and Elijah, talking with Jesus**.*
>
> *Peter said to Jesus, "Lord, it is good for us to be here. If you wish, I will put up three shelters—one for you, one for Moses and one for Elijah."*
>
> *While he was still speaking, a bright cloud enveloped them, and a voice from the cloud said, "**This is my Son, whom I love; with him I am well pleased. Listen to him**!"*
>
> *When the disciples heard this, they fell facedown to the ground, terrified.*
>
> *But Jesus came and touched them. "Get up," he said. "Don't be afraid." **When they looked up, they saw no one except Jesus.** (Matthew [17:1–7])*

The promises initiated in the Old Testament and pictured by God's holy days ultimately will find fulfillment in the New Testament by the better means of Christ, as he replaces the fallible and temporary lambs and bloodletting types of the old.

Christ died at the start of the New Testament, but the church, the rest of the firstfruits of the harvest, continued on. They continued to meet on the Sabbath, God's eternal weekday covenant, which is shown being kept again in the future after God creates the new heavens and earth.

> *Wherefore the children of Israel shall keep the sabbath,* to observe the sabbath throughout their generations, *for a perpetual covenant. It is a sign between me and the children of Israel for ever*: for in six days the LORD made heaven and earth, and on the seventh day he rested, and was refreshed. (Exodus 31:16–17, KJV)

> For *as the new heavens and the new earth,* which I will make, shall remain before me, saith the LORD, so shall your seed and your name remain.
> And *it shall come to pass,* that **from one new moon to another,** *and from one sabbath* **to another,** shall all flesh come to worship before me, saith the LORD. (Isaiah 66:22–23, KJV)

The New Testament church was demanded to change from the eternal Sabbath covenant to keeping the first day of the week by the order of a Roman emperor hundreds of years after AD 31. I don't need to remind you, there is a power that seeks to change days and times we are warned about. Notwithstanding this submission, it apparently gets changed back in the new Jerusalem of the new heavens, as the temple once again gets swept clean from paganism.

Figure 7. The Edict of Constantine 321 AD (http:amazingdiscoveries.org/S-deception-Sabbath_change_Constantine)

Sunday actually made very little headway as a Christian day of rest until the time of Constantine in the fourth century. Constantine was emperor of Rome from AD 306 to 337. He was a sun worshiper during the first years of his reign. Later, he professed conversion to Christianity, but at heart remained a devotee of the sun. Edward Gibbon says, "The Sun was universally celebrated as the invincible guide and protector of Constantine."[ii]

Constantine created the earliest Sunday law known to history in AD 321. It says this:

> On the venerable Day of the sun let the magistrates and people residing in cities rest, and let all workshops be closed. In the country, however, persons engaged in agriculture may freely and lawfully continue their pursuits: because it often happens that another Day is not so suitable for grain sowing or for vine planting: lest by neglecting the proper moment for such operations the bounty of heaven should be lost.[ii]

Constantine's Conversion by Peter Paul Rubens
Source: Wikimedia Commons....

Chamber's Encyclopedia says this:

> Unquestionably the first law, either ecclesiastical or civil, by which the Sabbatical observance of that Day is known to have been ordained, is the edict of Constantine, 321 A.D.[iii]

Following this initial legislation, both emperors and Popes in succeeding centuries added other laws to strengthen Sunday observance. What began as a pagan ordinance ended as a Christian regulation. Close on the heels of the Edict of Constantine followed the Catholic Church Council of Laodicea (circa 364 AD):

> Christians shall not Judaize and be idle on Saturday (Sabbath), but shall work on that Day: but the Lord's Day, they shall especially honour; and as being Christians, shall, if possible, do no work on that day. If however, they are found Judaizing, they shall be shut out from Christ.[iv]

Keeping the day of the sun, yuletide logs that represent the sun's rebirth, evergreen trees, fertile bunny rabbits, which now choke the temple, seem such a chosen blindness. They are such a poor deception compared to the brilliant, thought-provoking days from God. Like paganism always has, it leads us away from what God intended, and away from what God shows us, in His Word, the Bible. But we are blind to that, aren't we? Hopefully only in ignorance and not knowingly.

From AD 31 until AD 321, Christians kept the weekly Sabbath Saturday until a Roman emperor changed the days. The change in modern times for the Worldwide Church of God (WCG) had a very old requirement beyond grace. This Sunday-keeping was not an option; it was demanded and was seemingly more important than our precious newfound grace. If we were slow on the uptake of Constantine's edict, we were vetoed and had to leave.

Please stop and reason for a minute. Do you not find it incredulous the churches of God on earth today generally observe sun-worshipping attributes? Sunday weekly meetings, and of course, everyone's favorite, the winter solstice, where the sun is attributed new life, as days get longer— Christmas. And Easter, well at least it is close to the correct date of the intended Passover. But where does the name come from? Again, I find it incredulous that the word *Easter* is not in the word of God, the Bible. Did you know that? It is a blatant and shameful mistranslation found only in the King James Version.

> *And because he saw it pleased the Jews, he proceeded further to take Peter also. (Then were the days of unleavened bread.)*
> *And when he had apprehended him, he put him in prison, and delivered him to four quaternions of soldiers to keep him; intending after **Easter** to bring him forth to the people. (Acts 12:3–4, KJV)*

Figure 8. Passover-not Easter (International Scripture Analyzer ISA)

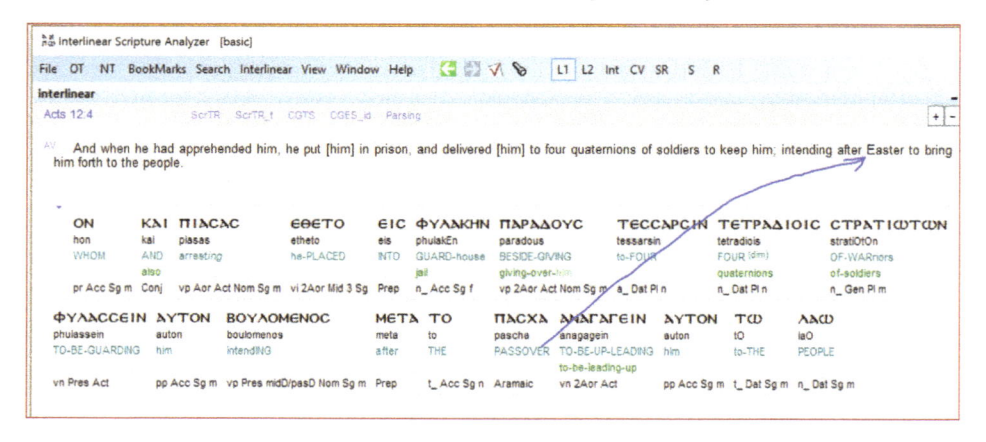

Who changed that? Who is driving the deception? Beware!

*About that time King Herod cruelly attacked some who belonged to the church, and he killed James, John's brother, with the sword. When he saw that it pleased the Jews, he proceeded to arrest Peter too, during the days of Unleavened Bread. After the arrest, he put him in prison and assigned four squads of four soldiers each to guard him, intending to bring him out to the people after the **Passover**. So Peter was kept in prison, but prayer was being made earnestly to God for him by the church. (Acts 12:1–5, HCSB)*

*I warn everyone who hears the words of the prophecy of this book: **If anyone adds anything to them, God will add to him the plagues described in this book**. (Revelation 22:18, NIV)*

For such are false apostles, deceitful workers, transforming themselves into the apostles of Christ.
*And no marvel; **for Satan himself is transformed into an angel of light**.*
Therefore it is no great thing if his ministers also be transformed as the ministers of righteousness; whose end shall be according to their works. (2 Corinthians 11:13–15, KJV)

*The great dragon was hurled down—that ancient serpent called **the devil, or Satan, who leads the whole world astray**. He was hurled to the earth, and his angels with him. (Revelation 12:9, NIV)*

***And he shall speak great words against the most High, and shall wear out the saints of the most High, and think to change times and laws**: and they shall be given into his hand until a time and times and the dividing of time. (Daniel 7:25, KJV)*

Well, so what? Or when is grace not enough?

Why am I pointing this out? Well, because I think it is true for everyone and also it is a rebuttal I didn't realize I was writing. I didn't know what I was doing until I see it now! Oh, I wrote it, but it is like I am just the pen and God is moving my hand. I had no plan to do this. It just happened by the will of God. This is a rebuttal to the events that savaged the Worldwide Church of God.

Around 2006, the church council of the Calgary, Alberta Congregation of the Worldwide Church of God had a vote. The vote was about how to change. And I agree we needed to. Without realizing it, we had been too close to living under the old covenant, where, if we did our work properly, we could earn salvation. So we would change and join into the greater wave of grace, but how much? What additional changes, beyond growing in grace, did we need to make?

The members of the council, myself included, voted to respect the "incarnation of Christ," even at the obviously wrong date of Christmas, and to recognize the shadows of the realities, respecting more the New Testament fulfillments on the anniversary dates of the ancient holy days of God as outlined from the ancient shadow days timing. We correctly wanted to adjust to the realization our works don't cut it. We are living in the grace of the new covenant. And we realized, there are many honest, God-loving people in other congregations. Before then, only members of WCG were true Christians.

That was never true, and I am glad we can now rejoice within the greater brotherhood. (I personally do not agree with everything they think, but grace makes them 1,000 percent God's people). I am thankful for that part of the change. I am now certain God has many sons and daughters in all churches, and He will be the judge of who they are!

But the problems and the reason for my rebuttal are the additional impurities this greater body brings into the temple we would be forced to adopt. For me, we would be listening to the voice responsible for the changing of days and times to pagan days and times and, especially for me, the untruth of the Trinity Godhead.

So the vote was about how to accept this adoption into the greater grace-filled body of Christ yet maintain some of our preferences about what God had shown our unique branch. To us, we still saw a better connection from the old covenant to the new covenant.

Like the ancient foreshadowing of the manna, the Bread that came down from heaven, is fulfilled in a better and more complete way with the true Bread of life—Jesus.

We are not searching on the ground for the food of angels. We, now in the New Testament, have a better bread. So in similar fashion, the other types, like the feast days of God, have us looking for the better fulfillment realities in the new.

Learn The Bible

Home » Bible » 1 Peter

Christ the True Manna

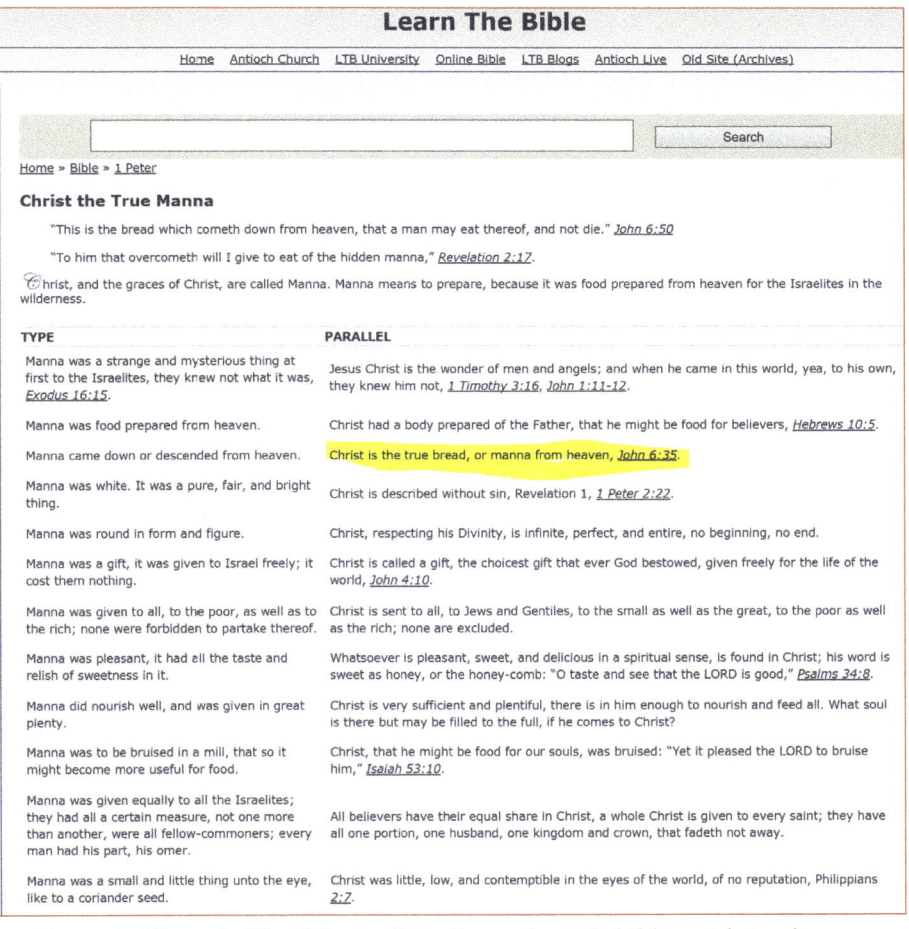

"This is the bread which cometh down from heaven, that a man may eat thereof, and not die." John 6:50

"To him that overcometh will I give to eat of the hidden manna," Revelation 2:17.

Christ, and the graces of Christ, are called Manna. Manna means to prepare, because it was food prepared from heaven for the Israelites in the wilderness.

TYPE	PARALLEL
Manna was a strange and mysterious thing at first to the Israelites, they knew not what it was, Exodus 16:15.	Jesus Christ is the wonder of men and angels; and when he came in this world, yea, to his own, they knew him not, 1 Timothy 3:16, John 1:11-12.
Manna was food prepared from heaven.	Christ had a body prepared of the Father, that he might be food for believers, Hebrews 10:5.
Manna came down or descended from heaven.	Christ is the true bread, or manna from heaven, John 6:35.
Manna was white. It was a pure, fair, and bright thing.	Christ is described without sin, Revelation 1, 1 Peter 2:22.
Manna was round in form and figure.	Christ, respecting his Divinity, is infinite, perfect, and entire, no beginning, no end.
Manna was a gift, it was given to Israel freely; it cost them nothing.	Christ is called a gift, the choicest gift that ever God bestowed, given freely for the life of the world, John 4:10.
Manna was given to all, to the poor, as well as to the rich; none were forbidden to partake thereof.	Christ is sent to all, to Jews and Gentiles, to the small as well as the great, to the poor as well as the rich; none are excluded.
Manna was pleasant, it had all the taste and relish of sweetness in it.	Whatsoever is pleasant, sweet, and delicious in a spiritual sense, is found in Christ; his word is sweet as honey, or the honey-comb: "O taste and see that the LORD is good," Psalms 34:8.
Manna did nourish well, and was given in great plenty.	Christ is very sufficient and plentiful, there is in him enough to nourish and feed all. What soul is there but may be filled to the full, if he comes to Christ?
Manna was to be bruised in a mill, that so it might become more useful for food.	Christ, that he might be food for our souls, was bruised: "Yet it pleased the LORD to bruise him," Isaiah 53:10.
Manna was given equally to all the Israelites; they had all a certain measure, not one more than another, were all fellow-commoners; every man had his part, his omer.	All believers have their equal share in Christ, a whole Christ is given to every saint; they have all one portion, one husband, one kingdom and crown, that fadeth not away.
Manna was a small and little thing unto the eye, like to a coriander seed.	Christ was little, low, and contemptible in the eyes of the world, of no reputation, Philippians 2:7.

Figure 9. Christ the True Manna (http://www.learnthebible.org/christ-the-true-manna.html)

The old was not a mistake to cast out in disgust. The mankind component had failed, and God had always known that would happen. So He gave us the spiritual way to get to the same promises—Christ. And just like the physical is a type, that eventually gets changed to the spiritual reality. We saw God's feast days in a new light and didn't want to throw what they teach away.

So here is where grace was not enough. We could not stay part of them without rejecting our personal preference on worship days. We had to change for these men but not for God!

> *Therefore **do not let anyone judge you** by what you eat or drink, or **with regard to a religious festival**, a New Moon celebration or a Sabbath day. **These are a shadow of the things that were to come; the reality, however, is found in Christ.** (Colossians 2:16–17, NIV)*

For example, in the old covenant, there was the wave sheaf—singular, special, first of the firstfruits, while in the new covenant, Jesus is what the wave sheaf was all about—first to be born and master of the Spirit harvest pictured in the parable of the sower and only begotten of the sons (firstfruits).

From the old, there was the early-spring barley harvest during the Days of Unleavened Bread and the greater fall wheat harvest called the Feast of Tabernacles. To my eyes, these physical harvests picture/foreshadow the future spiritual harvests (old/new, physical/spiritual).

In the new, there is Christ resurrected (premium wave sheaf), then the first harvest (barley = kings and priests) of the saints at the return of Christ, and later, as spoken of in revelation, after a thousand years of the kingdoms reign, "the rest of the dead" (honorable wheat harvest) are in a much larger and complete resurrection of those (still asleep) who died not knowing their God yet.

> *For the Lord himself will come down from heaven, with a loud command, with the voice of the archangel and with the trumpet call of God, and the dead in Christ*

> *will rise first. After that, we who are still alive and are left will be caught up together with them in the clouds to meet the Lord in the air. And so we will be with the Lord forever. (1 Thessalonians 4:16–17, NIV)*
>
> *I saw thrones on which were seated those who had been given authority to judge. And I saw the souls of those who had been beheaded because of their testimony for Jesus and because of the word of God. They had not worshiped the beast or his image and had not received his mark on their foreheads or their hands. They came to life and reigned with Christ a thousand years. (The rest of the dead did not come to life until the thousand years were ended.) This is the first resurrection. Blessed and holy are those who have part in the first resurrection.* **The second death has no power over them***, but they will be priests of God and of Christ and will reign with him for a thousand years. (Revelation 20:4–6, NIV)*

This world, for a thousand years, having no influence from Satan, the prince of the power of the air, will show a flabbergasting improvement. The people of the great resurrection (the second, formerly foreshadowed as the wheat harvest) will probably hardly recognize life on earth after all that healing.

However, lest we think we could be good, Satan is loosed and drives the world to madness again for a short time. After that is corrected, every living being will know without a doubt **only God can be good**.[15]

So to recap a bit, we felt strongly that the means to the promises are typified by the spiritual fulfillment of what the feasts of God pictured. And that they contain much better teachings and ideas to focus on, in contrast to the attributes of paganism found in yuletide logs, the winter solstice, bunny rabbits and hot cross buns, and the other dressings of paganism (that are, I think, are a stench to God) that this new greater graceful body insists on keeping in Gods temple (the church).

We wanted to join this greater body but maintain some of the cleanliness God had put in our hearts and minds. We did not disrespect them in this but wanted to bring this view with us in the greater temple.

The plan of God and the means to successfully inheriting what God promised man and the joining in His family has transcended and been a component of both covenants. There is only one Israel, and Christians and gentiles are either grafted into it or reintroduced to it as actual sons of Abraham. There is only the new Jerusalem and only the new temple in Israel eventually, and we are all sons of Abraham.

At the least, we consider the old covenant now as something like when Christ walks across a stage, His shadow still follows. He has his shadow, and although it is only that (a shadow), it is forever with Him as part of our history.

Like old Jerusalem, old Israel, the promised greatest nation and company of nations are still there today. *Genesis 35:11 (NIV) "And God said to him, 'I am God Almighty; be fruitful and increase in number. A nation and a community of nations will come from you, and kings will come from your body.'"*

But the physical old covenant is passing, and its days as even a remnant are numbered, because there will come a time when there is no day and night. But until then, let's remember them on the anniversaries and especially honoring the body and not the shadow.

Again:

> *Therefore do not let anyone judge you by what you eat or drink, or with regard to a religious festival, a New Moon celebration or a Sabbath day.* ***These are a shadow of the things that were to come; the reality, however, is found in Christ.*** *(Colossians 2:16–17, NIV)*

We planned that Christmas would be the "Incarnation Day," Easter would be Passover, and the other holy days would be recognized as Old Testament but still honored for their association

==with their New Testament spiritual fulfillment in Christ== *==(which includes Passover, the wave sheaf, Pentecost, trumpets, atonement, and the harvests of God's people)==.*

==We presented that decision to the pastor, and he vetoed it, stating it would not suit the direction from the voices with authority on high. Like most churches, we must newly accept wholeheartedly the changes of times and seasons, as ordained long ago by the authority of the Roman emperor and the Roman Catholic Church. I didn't like that, and my wife and I left (along with some friends). I still love and respect that pastor because I know he is a man of God, but I think that was like building an Asherah pole in the temple of God and a big mistake.==

My heart is pouring out as I write this. I feel like Judah of the dispersion as we were forced into Babylon. The temple was torn down and desecrated. But thankfully, I know it will be rebuilt. That is happening now by the powerful, unstoppable Spirit of God. What God says in the Bible, it shall come to pass. Sometimes, it seems we have misplaced the Bible, but like King Josiah and the high priest Hilkiah, the book is still in the temple—we just need to read and heed! Thankfully, although we have forgotten God, He has not forgotten us, and He is restoring His eternal Israel (us). We too shall keep a proper Passover in the future.

I was told to stop talking about the Passover so much. Wow, sorry, but I will expound the word of God still and even more than before if it is God's will. *Luke 22:15 (NIV): "And he said to them, 'I have eagerly desired to **eat this Passover with you before I suffer**. For I tell you, **I will not eat it <u>again</u>** until it finds fulfillment in the kingdom of God.'"*

Christ is our Passover, and He will celebrate Passover in the future kingdom of God. So says the Lord. You can change that, if you like, and call it the Lord's Supper, but that is not what Christ calls it. I believe we have a future date with the actual Passover Lamb that we will reflect back upon, on the appropriate dates of its historical inception, basking much better in the glorious New Testament fulfillment that is Christ. Not one single commentary I reviewed would accept

Christ at His word on this scripture. How about you? Will you eat at this Passover meal with Christ in the future kingdom of God?

Now back from my holy day digression and rebuttal to the changes forced against the members of the Worldwide Church of God.

Again, who was the first to be resurrected from the dead? The apostle Paul said "that the Christ would suffer, that He would be the *first to rise from the dead,* and would proclaim light to the Jewish people and to the Gentiles" (Acts 26:23). Jesus was the first of God's spiritual harvest to be resurrected (see also Colossians 1:13–18), and He had become the first (firstborn) or the first of the firstfruits of those who have died.

> *He is the image of the invisible God,* **the firstborn** *over all creation. For by him all things were created: things in heaven and on earth, visible and invisible, whether thrones or powers or rulers or authorities; all things were created by him and for him. He is before all things, and in him all things hold together. And he is the head of the body, the church; he is the beginning* **and the firstborn from among the dead**, *so that in everything he might have the supremacy. (Colossians 1:15–18, NIV)*

Like the wave sheaf was presented singularly, so was Christ isolated for the harvest **presented singularly** as the first resurrected into the spiritual kingdom of God.

> *For the Lord himself will come down from heaven, with a loud command, with the voice of the archangel and* **with the trumpet call of God**, *and the dead in Christ will rise first. (1 Thessalonians 4:16, NIV)*

> *Listen, I tell you* **a mystery***: We will not all sleep, but* **we will all be changed**— *[52] in a flash,* **in the twinkling of an eye, at the last trumpet***. For the trumpet*

> *will sound, the dead will be raised imperishable, and we will be changed. (1 Corinthians 15:51–52, NIV)*
>
> *Dear friends, now we are children of God, **and what we will be has not yet been made known**. ==But we know that when he appears, we shall be like him, for we shall see him as he is.== (1 John 3:2, NIV)*

We too will then be spiritually at one with Jesus and the Father. *John 17:21 (NIV): "That all of them may be one, Father, just as you are in me and I am in you. May they also be in us so that the world may believe that you have sent me."*

So we see, the means to the promises found in the Old Testament (God's holy days were only a physical foreshadowing of the real things) are now in a spiritual way the means to the new covenant–kingdom promises. There is still a Passover (wave sheaf), firstfruits, harvest, Israel, Trumpet call, Atonement, new Jerusalem, new temple, etc., and they are much greater than pagan counterfeits generally adopted today).

> *And that from a child thou hast **known the holy scriptures, which are able to make thee wise unto salvation through faith which is in Christ Jesus.**
> **All scripture is given by inspiration of God**, and is profitable for doctrine, for reproof, for correction, for instruction in righteousness:*
> ***That the man of God may be perfect,***
> ***throughly furnished unto all good works.***
> *(2 Timothy 3:15–17, KJV)*

Now let us consider those scriptures and see about our future in space.

The words below are somewhat dual and are both for ancient Israel of the old covenant and the future Israel of the new. There should be trumpets blaring at this announcement of the Creator! He is coming to make it so. It shows how great the contrast is between the glorious greatness of God, who inhabits space, and the insignificance of physical mankind.

However, that is going to change. What an inheritance! We will share in Christ's glory!

There was never anything like this showered on physical Israel. It is for God's family, and it is so great it is almost "unseeable," like "we can't know."

So I hope with renewed and appropriate reverence, we can consider these words in a new light, as His plan **for all time**. As you read, may God restore your joy of salvation and may we begin to see how great it is.

This is our Creator speaking about our salvation—future— through the mouth of the holy prophets. The plan for the salvation of Israel-mankind transcends both covenants. The old, though transcended, is the path to the new. Together, **old** and **new**, **the word** of God paints our future. The promises still apply. Don't ignore the Old Testament. Both testaments sing of it.

> *They sang the song of God's servant* ***Moses and*** *the song of* ***the Lamb***:
> *Great and awe-inspiring are Your works,*
> *Lord God, the Almighty;*
> *righteous and true are Your ways, King of the Nations.*
> *(Revelation 15:3, HCSB)*
>
> ***And he shall send Jesus Christ, which before was preached unto you:***
> ***Whom the heaven must receive until the times of restitution of all things,*** ==**which God hath spoken by the mouth of all his holy prophets**== ***since the world*** ==**began**==.
> *For Moses truly said unto the fathers, A prophet shall the Lord your God raise up unto you of your brethren, like unto me; him shall ye* **hear in all things whatsoever he shall say** *unto you. (Acts 3:20–22, KJV)*

> *He says*:
> **"It is too small a thing for you to be my servant**
> **to restore the tribes of Jacob**
> *and bring back those of Israel I have kept.*
> **I will also make you a light for the Gentiles,**
> ==**that you may bring my salvation to the ends of the earth."**== *(Isaiah 49:6)*
>
> *Now listen*:
> *You will conceive and give birth to a son,*
> *and you will call His name* ==**Jesus.**==
> *He will be great*
> *and will be called the Son of the Most High,*
> *and the Lord God will give Him*
> **the throne of His father David.**
> ==**He will reign over the house of Jacob forever,**==
> ==**and His kingdom will have no end.**==
> *(Luke 1:31–33, HCSB)*

Jesus, in the future and with us, will rule over Israel. Israel is now composed of wild olive branches (gentiles) and the returning natural ones (covering all mankind).

Remember, the first order of New Testament business was for the apostles to go to the lost sheep of Israel. The gentiles and Samaritans would be welcomed later. Israel was not forgotten nor forsaken.

> *These twelve Jesus sent out with the following instructions: "Do not go among the Gentiles or enter any town of the Samaritans.* **Go rather to the lost sheep of Israel**. *As you go, preach this message: 'The kingdom of heaven is near.'" (Matthew 10:5–7, NIV)*

The **transcending root**, which is Christ, sustains the continuing and entire olive plant—all Israel (future rulers with God). *(More on this elsewhere, but I must interject. Christ is the "I am" and, therefore,* *was the God of the old testament, so He has always been the sustainer of the natural olive plant, to which he is now adding wild olive branches and restoring believing natural branches (all Israel). The olive plant, was never destroyed; it was cleansed of the unbelieving, for which God has now provided a new and better way to it! See* "I Am that I Am" *on p. 146.)*

What then? Israel did not find what it was looking for, but the elect did find it. The rest were hardened, as it is written:
> **God gave them a spirit of insensitivity,**
> **eyes that cannot see**
> **and ears that cannot hear, to this day.**

And David says:
> **Let their feasting become a snare and a trap,**
> **a pitfall and a retribution to them.**
> **Let their eyes be darkened so they cannot see,**
> **and their backs be bent continually.**

==*Israel's Rejection Not Final*==
I ask, then, have they stumbled in order to fall? Absolutely not! On the contrary, by their stumbling, salvation has come to the Gentiles to make Israel jealous. Now if their stumbling brings riches for the world, and their failure riches for the Gentiles, how much more will their full number bring!

Now I am speaking to you Gentiles. In view of the fact that I am an apostle to the Gentiles, I magnify my ministry, if I can somehow make my own people jealous and save some of them. For if their rejection brings reconciliation to the world, what will their acceptance mean but life from the dead? ==**Now if the firstfruits offered up are holy, so is the whole batch. And if the root is holy, so are the branches.**==

Now if some of the branches were broken off, and you, though a wild olive branch, were grafted in among them and have come to share in the rich root of the cultivated olive tree, do not brag that you are better than those branches. But if you do brag—you do not sustain the root, but the root sustains you. Then you will say, "Branches were broken off so that I might be grafted in." True enough; they were broken off by unbelief, but you stand by faith. Do not be arrogant, but be afraid. For if God did not spare the natural branches, He will not spare you either. Therefore, consider God's kindness and severity: severity toward those who have fallen but God's kindness toward you—if you remain in His kindness. Otherwise you too will be cut off. And even they, ==**if they do not remain in unbelief, will be grafted in, because God has the power to graft them in again.**== *For if you were cut off from your native wild olive and against nature were grafted into a cultivated olive tree,* ==*how much more will these—the natural branches—be grafted into their own olive tree?*==

So that you will not be conceited, brothers, I do not want you to be unaware of this mystery: ==*A partial hardening has come to Israel until the full number of the Gentiles has come in.*== ==**And in this way all Israel will be saved, as it is written:**==

    The Liberator will come from Zion;
    He will turn away godlessness from Jacob.
    And this will be My covenant with them
    when I take away their sins.

Regarding the gospel, they are enemies for your advantage, but regarding election, they are loved because of the patriarchs, since God's gracious gifts and calling are irrevocable. As you once disobeyed God, but now have received mercy through their disobedience, so they too have now disobeyed, resulting in mercy to you, so that they also now may receive mercy. For God has imprisoned all in disobedience, so that He may have mercy on all.

A Hymn of Praise
*Oh, the depth of the riches
both of the wisdom and the knowledge of God!
How unsearchable His judgments
and untraceable His ways!*
**For who has known the mind of the Lord?
Or who has been His counselor?
Or who has ever first given to Him,
and has to be repaid?**
*For from Him and through Him
and to Him are all things.
To Him be the glory forever. Amen.
(Romans 11:7–36, HCSB)*

<u>**When the Messiah, who is your life, is revealed**</u>, <u>*then*</u> *you also will be revealed with Him in glory. (Colossians 3:4, HCSB)*

Therefore, as a fellow elder and witness to the sufferings of the Messiah and **also a participant in the glory about to be revealed**, *I exhort the elders among you. (1 Peter 5:1, HCSB)*

Comfort for God's People
<u>**Comfort, comfort my people,
says your God.
Speak tenderly to Jerusalem,
and proclaim to her**</u>
*that her hard service has been completed,
that her sin has been paid for,
that she has received from the LORD's hand*

double for all her sins.
A voice of one calling:
"In the desert prepare
the way for the LORD;
make straight in the wilderness
a highway for our God.
Every valley shall be raised up,
every mountain and hill made low;
the rough ground shall become level,
the rugged places a plain.
==**And the glory of the LORD will be revealed,**==
==**and all mankind together will see it.**==
==**For the mouth of the LORD has spoken."**==
A voice says, "Cry out."
And I said, "What shall I cry?"
"All men are like grass,
and all their glory is like the flowers of the field.
The grass withers and the flowers fall,
because the breath of the LORD blows on them.
Surely the people are grass.
The grass withers and the flowers fall,
*but **the word of our God stands forever**."*
You who bring good tidings to Zion,
go up on a high mountain.
You who bring good tidings to Jerusalem,
lift up your voice with a shout,
lift it up, do not be afraid;
say to the towns of Judah,
"Here is your God!"
See, the Sovereign LORD comes with power,
and his arm rules for him.
==**See, his reward is with him,**==
and his recompense accompanies him.
==**He tends his flock like a shepherd:**==
He gathers the lambs in his arms
and carries them close to his heart;

he gently leads those that have young.
Who has measured the waters in the hollow of his hand,
or with the breadth of his hand marked off the heavens?
Who has held the dust of the earth in a basket,
or weighed the mountains on the scales
and the hills in a balance?
==*Who has understood the mind of the LORD,*==
==*or instructed him as his counselor?*== [Isaiah 40:13 of the *New American Standard Exhaustive Concordance:* "Who has directed the Spirit of the LORD? Or as His counsel or has informed Him?"]
Whom did the LORD consult to enlighten him,
and who taught him the right way?
Who was it that taught him knowledge
or showed him the path of understanding?
Surely the nations are like a drop in a bucket;
they are regarded as dust on the scales;
he weighs the islands as though they were fine dust.
Lebanon is not sufficient for altar fires,
nor its animals enough for burnt offerings.
Before him all the nations are as nothing;
they are regarded by him as worthless
and less than nothing.
To whom, then, will you compare God?
What image will you compare him to?
As for an idol, a craftsman casts it,
and a goldsmith overlays it with gold
and fashions silver chains for it.
A man too poor to present such an offering
selects wood that will not rot.
He looks for a skilled craftsman
to set up an idol that will not topple.
Do you not know?
Have you not heard?
Has it not been told you from the beginning?
Have you not understood since the earth was founded?

He sits enthroned above the circle of the earth.
[Notice verse 22 "above the circle of the earth" (or vault). Worth considering that this ancient book shows early knowledge, that the earth was not flat, but a sphere?]
and its people are like grasshoppers.
He stretches out the heavens like a canopy,
and spreads them out like a tent to live in.
He brings princes to naught
and reduces the rulers of this world to nothing.
No sooner are they planted,
no sooner are they sown,
no sooner do they take root in the ground,
than he blows on them and they wither,
and a whirlwind sweeps them away like chaff.
"To whom will you compare me?
Or who is my equal?" says the Holy One.
Lift your eyes and look to the heavens:
Who created all these?
He who brings out the starry host one by one,
and calls them each by name.
Because of his great power and mighty strength,
not one of them is missing.
Why do you say, **O Jacob**,
and complain, **O Israel**,
"My way is hidden from the LORD;
my cause is disregarded by my God"?
Do you not know?
Have you not heard?
The LORD is the everlasting God,
the Creator of the ends of the earth.
He will not grow tired or weary,
and his understanding no one can fathom.
He gives strength to the weary
and increases the power of the weak.
Even youths grow tired and weary,

and young men stumble and fall;
but those who hope in the LORD
will renew their strength.
They will soar on wings like eagles;
they will run and not grow weary,
they will walk and not be faint. (Isaiah 40)

Yes, God describes our future venue, space, but who has seen it? God made all the stars and the ends of the earth. That is where God exists (everywhere and everywhen), above the "circle" of the earth. And that is what he brings to us through the efforts of our Messiah. **He will save his people Israel**, and it will include the wild olive branches of the gentiles. **For He makes us all children of Abraham, and this glorious space is the promise we shall inherit.**

Perhaps you think the Old Testament words of Isaiah should be forgotten and are not for Christians? Please reconsider.

For to us a child is born,
to us a son is given,
and the government will be on his shoulders.
And he will be called
Wonderful Counselor, Mighty God,
Everlasting Father, Prince of Peace.
Of the increase of his government and peace
there will be no end.
He will reign on David's throne
and over his kingdom,
establishing and upholding it
with justice and righteousness
from that time on and forever.
The zeal of the LORD Almighty
will accomplish this. (Isaiah 9:6–7, NIV)

He said to me, "You are my servant,
Israel, in whom I will display my splendor."

SPACE THE TRUE FRONTIER!

But I said, "I have labored to no purpose;
I have spent my strength in vain and for nothing.
Yet what is due me is in the LORD's hand,
and my reward is with my God."
And now the LORD says—
he who formed me in the womb to be his servant
to bring Jacob back to him
and gather Israel to himself,
for I am honored in the eyes of the LORD
and my God has been my strength—
he says:
"It is too small a thing for you to be my servant
to restore the tribes of Jacob
and bring back those of Israel I have kept.
I will also make you a light for the Gentiles,
that you may bring my salvation to the ends of the
earth" **[Romans 11:26–27].**
This is what the LORD says—
the Redeemer and Holy One of Israel—
to him who was despised and abhorred by the nation,
to the servant of rulers:
"Kings will see you and rise up,
princes will see and bow down,
because of the LORD, who is faithful,
the Holy One of Israel, who has chosen you."
Restoration of Israel
This is what the LORD says:
"In the time of my favor I will answer you,
and in the day of salvation I will help you;
I will keep you and will make you
to be a covenant for the people,
to restore the land
and to reassign its desolate inheritances,
to say to the captives, 'Come out,'
and to those in darkness, 'Be free!'
"They will feed beside the roads

and find pasture on every barren hill.
They will neither hunger nor thirst,
nor will the desert heat or the sun beat upon them.
He who has compassion on them will guide them
and lead them beside springs of water.
I will turn all my mountains into roads,
and my highways will be raised up.
See, they will come from afar—
some from the north, some from the west,
some from the region of Aswan."
Shout for joy, O heavens;
rejoice, O earth;
burst into song, O mountains!
For the LORD comforts his people
and will have compassion on his afflicted ones.
But Zion said, "The LORD has forsaken me,
the Lord has forgotten me."
"Can a mother forget the baby at her breast
and have no compassion on the child she has borne?
Though she may forget,
I will not forget you!
See, I have engraved you on the palms of my hands;
your walls are ever before me.
Your sons hasten back,
and those who laid you waste depart from you.
Lift up your eyes and look around;
all your sons gather and come to you.
As surely as I live," declares the LORD,
"you will wear them all as ornaments;
you will put them on, like a bride.
"Though you were ruined and made desolate
and your land laid waste,
now you will be too small for your people,
and those who devoured you will be far away.
The children born during your bereavement

SPACE THE TRUE FRONTIER!

will yet say in your hearing,
'This place is too small for us;
give us more space to live in.'
Then you will say in your heart,
'Who bore me these?
I was bereaved and barren;
I was exiled and rejected.
Who brought these up?
I was left all alone,
but these—where have they come from?'"
This is what the Sovereign LORD says:
"See, I will beckon to the Gentiles,
I will lift up my banner to the peoples;
they will bring your sons in their arms
and carry your daughters on their shoulders.
Kings will be your foster fathers,
and their queens your nursing mothers.
They will bow down before you with their faces to the ground;
they will lick the dust at your feet.
Then you will know that I am the LORD;
those who hope in me will not be disappointed."
Can plunder be taken from warriors,
or captives rescued from the fierce?
But this is what the LORD says:
"Yes, captives will be taken from warriors,
and plunder retrieved from the fierce;
I will contend with those who contend with you,
and your children I will save.
I will make your oppressors eat their own flesh;
they will be drunk on their own blood, as with wine.
Then all mankind will know
that I, the LORD, am your Savior,
your Redeemer, the Mighty One of Jacob."
(Isaiah 49:3–26, NIV)

Everlasting Salvation for Zion
"Listen to me, you who pursue righteousness
and who seek the LORD:
**Look to the rock from which you were cut
and to the quarry from which you were hewn;
look to Abraham, your father,**
and to Sarah, who gave you birth.
When I called him he was but one,
and I blessed him and made him many.
The LORD will surely comfort Zion
and will look with compassion on all her ruins;
he will make her deserts like Eden,
her wastelands like the garden of the LORD.
Joy and gladness will be found in her,
thanksgiving and the sound of singing.
"Listen to me, my people;
hear me, my nation:
The law will go out from me;
my justice will become a light to the nations.
**My righteousness draws near speedily,
my salvation is on the way,**
and my arm will bring justice to the nations.
The islands will look to me
and wait in hope for my arm.
Lift up your eyes to the heavens,
look at the earth beneath;
the heavens will vanish like smoke,
the earth will wear out like a garment
and its inhabitants die like flies.
**But my salvation will last forever,
my righteousness will never fail.** *(Isaiah 51:1–6)*

This is for people of all times. It involves our true salvation as eternal beings at one with God. It crosses the covenants and is shown below in the New Testament, that all Christian's are part of these original promises God made to Abraham.

God didn't start over. He didn't fail. What God says goes! Always and forever. Amen.

> *You are all sons of God through faith in Christ Jesus, for all of you who were baptized into Christ have clothed yourselves with Christ.* **There is neither Jew nor Greek, slave nor free, male nor female, for you are all one in Christ Jesus.** <mark>**If you belong to Christ, then you are Abraham's seed, and heirs according to the promise.**</mark> *(Galatians 3:26–29, NIV)*

As mankind under the old covenant, we failed in earning these promises. The fault was ours not Gods!

The old covenant relied on us keeping our part, but we could not, and no man could (except for one man, the God who would be man, the only begotten Son, Jesus).

The **new and better covenant is dependent on Christ**, and in him alone, we succeed and we are adopted into Gods family and inherit the universe prepared for us since before time!

What an awesome God. **The covenant was replaced but not the promises!**

> *What I mean is this: The law, introduced 430 years later,* ***does not set aside the covenant previously established by God and thus do away with the promise****. (Galatians 3:17, NIV)*

> *For if you were cut off from your native wild olive and against nature were grafted into a cultivated olive tree, how much more will these —* ***the natural branches*** *— be grafted into their own olive tree?*
> *So that you will not be conceited, brothers, I do not want you to be unaware of this mystery: A partial* **hardening has come to Israel until the full number of**

> *the Gentiles has come in. <u>And in this way all Israel will be saved, as it is written:</u>*
>   *The Liberator will come from Zion;*
>   *He will turn away godlessness from Jacob.*
>   *<u>And this will be My covenant with them</u>*
>   *<u>when I take away their sins.</u>*
> *Regarding the gospel, they are enemies for your advantage, but regarding election, they are loved because of the patriarchs, <u>since God's gracious gifts and calling are irrevocable</u>. As you once disobeyed God, but now have received mercy through their disobedience, so they too have now disobeyed, resulting in mercy to you, so that they also now may receive mercy. For God has imprisoned all in disobedience, so that He may have mercy on all. (Romans 11:24–32, HCSB)*

All are now one-in-Christ, the true olive root, and are sons of Abraham that will inherit the promises made to Israel (now spiritual, new covenant)

There is no new Messiah coming for Old Testament physical Israel (Christ was sacrificed once only). We are now all funneled through the Great "I Am," God of both testaments, both Jews and gentiles, man and woman. That is the only door (Vine = Christ) into the temple of God in New Jerusalem. Gods promise was for all mankind and transcends both covenants forever.

I look forward to the physical nation called Israel on earth today, being grafted back into the vine. It's the only door, and the only Messiah (Jesus), by which they will be saved. Amen

The fallen archangel, ex-light-bringer now Satan was initially the messenger in charge of helping mankind into the family of God. His job was to bring mankind into the glorious universe where God is.

We are to become one with God and eventually share in that glory—unified in spiritual existence and likeness to God (explanation and backup to follow). *Hebrews 1:14 (NIV): "Are not all angels ministering spirits sent to serve those who will inherit salvation?"*

SPACE THE TRUE FRONTIER!

That fallen angel is now completely anti-Christ. I speculate that Satan caused the current destructed state of the universe, him and perhaps as many as a third of the angels that chose to follow him and rebel against God.

> *He replied, "I saw Satan fall like lightning from heaven. (Luke 10:18, NIV)*

> *And there was war in heaven. Michael and his angels fought against the dragon, and the dragon and his angels fought back. But he was not strong enough, and they lost their place in heaven.* ***The great dragon was hurled down—that ancient serpent called the devil, or Satan, who leads the whole world astray. He was hurled to the earth, and his angels with him.*** *(Revelation 12:7–9, NIV)*

In Isaiah 14, the vanity stricken king of Babylon is compared to Satan (throughout the Old Testament, God flashes between the current physical situation with people and back to the same issue existing in the fallen Satan that preceded mankind).

It is also an interesting point that the word *Lucifer* in the King James Version is not in the original Hebrew source.[16]

Below is a picture taken from a free Internet program (Interlinear Scripture Analyser)[17] that shows the original words (Greek and Hebrew) and provides a word-by-word (interlinear) translation to English. It also shows the King James (Authorized Version) translation. I will be using this tool throughout this book.

As you can see from the original Hebrew, there was no Hebrew word relating to *Lucifer* and, like most other translations, should be translated "son of the morning" only.

Figure 10. Lucifer- no such name in Greek (International Scripture Analyzer ISA)

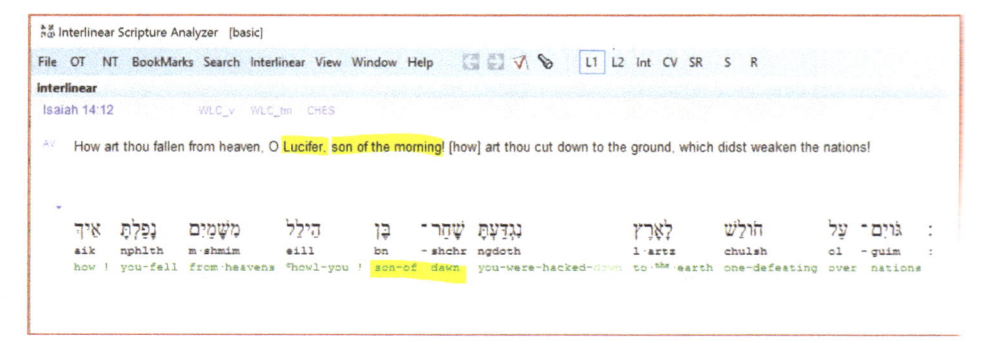

[Screenshot of Interlinear Scripture Analyzer showing Isaiah 14:12: "How art thou fallen from heaven, O Lucifer, son of the morning! [how] art thou cut down to the ground, which didst weaken the nations!" with Hebrew interlinear showing "how | you-fell from-heavens ^howl-you ! son-of dawn you-were-hacked-down to-the-earth one-defeating over nations"]

Later Christ takes over the role, or position (as this is a title and not a name), of the Light-Bringer, Son of Dawn, or Bright Morningstar, correctly fulfilling the leadership role that Satan fell away, or rebelled from, and was certainly disqualified from. The scary part, the almost unbelievable part, is that "we" Christians share with Christ in that role soon!

> ***And he who is overcoming, and who is keeping unto the end my works, I will give to him authority over the nations,*** *and he shall rule them with a rod of iron--as the vessels of the potter they shall be broken—* ***as I also have received from my Father; and I will give to him the morning star****. He who is having an ear--let him hear what the Spirit saith to the assemblies. (Revelation 2:26–29, YLT)*

> *"****Here's the reward*** *I have for every conqueror, everyone who keeps at it, refusing to give up:* ***You'll rule the nations****, your Shepherd-King rule as firm as an iron staff, their resistance fragile as clay pots. This was the gift my Father gave me;* ***I pass it along to you—[28] and with it, the Morning Star!****" (Revelation 2:26–28, MSG)*

The coming sons of God will rule with Christ (wow, I didn't say that, God did), sharing in the role as bright morning stars.

SPACE THE TRUE FRONTIER!

Below is a picture of the IVP Bible Background Commentary on Revelation 2:28–29 (by John H. Walton) that I find very interesting as a reference. Of course, today we understand Venus is a planet and not a star.

Figure 11. Sharing in Christ's rule (The IVP Bible Background Commentary Old and New Testaments)

> **The IVP Bible Background Commentary: Old and New Testaments** — 2:28-29
>
> **2:28-29.** The morning star, Venus, heralded the dawn, and great people could be compared to it as well as to the sun shining in glory (Sir 50:6); cf. Rev 22:16. Because most of the Greco-Roman world believed that life was ruled by the stars, to be given authority over one of the most powerful of stars (a symbol of sovereignty among the Romans) was to share *Christ's rule over creation (2:26-27).

*And we have the word of the prophets made more certain, and you will do well to pay attention to it, as to a light shining in a dark place, until the day dawns and **the morning star rises in your hearts**. (2 Peter **1:19,** NIV)*

Figure 12. The Messianic Star of the Morning, at The Lords Day (The IVP Bible Background Commentary Old and New Testaments)

> **The IVP Bible Background Commentary: Old and New Testaments** — 1:19
>
> **1:19.** The apostolic revelation in *Christ confirmed the revelations of the Old Testament prophets. Some *Dead Sea Scrolls texts present the "star" of Nu 24:17 as *messianic, and an Old Testament text describes the coming day of the Lord in terms of a sunrise (Mal 4:2) because God would come like the sun (cf. Ps 84:11). The point here seems to be that the morning star (Venus) heralds the advent of dawn; a new age was about to dawn (cf. 2Pe 1:11), but the Old Testament plus what was revealed by Jesus' first coming was the greatest revelation the world would experience until his return in the day of the Lord. "You do well" was a common way of suggesting that a person do something (i.e., "You *ought* to do this").

*"**I, Jesus**, have sent my angel to give you this testimony for the churches. I am the Root and the Offspring of David, and **the bright Morning Star**." (Revelation 22:16, NIV)*

And I am showing the above commentaries because I want to show the "morning star" is much more than just like the planet Venus but a star. (Also noteworthy, the Day of the Lord is not a day of the week).

More on sharing in the rule of God, a surprising revelation from Jesus, showing we shall drink from the same cup of Jesus.

> *Jesus responded, "You have no idea what you're asking." And he said to James and John, "**Are you capable of drinking the cup that I'm about to drink**?"*
> *They said, "**Sure**, why not?"*
> *Jesus said, "**Come to think of it, you are going to drink my cup.** But as to awarding places of honor, that's not my business. My Father is taking care of that." (Matthew 20:22–23, MSG)*

Again, this leadership role—bright morning star—with Jesus will include a kingdom that spreads to the vast universe (a new heaven and a new earth, Revelations 21:1–2).

Please note that Israel is not forgotten! The twelve tribes of Israel are something we will all be involved with in the future! (Sometimes, I think Christians, are the greatest doubters of the word of God. Who needs atheists to deny the Bible? Sometimes we do just fine. More and more, I find people in churches rarely use the Bible, preferring their own words. I have spoken with many Christians who hate the idea that Israel has anything to do with them. Yet look what God says.)

> *And **Jesus said** unto them, Verily I say unto you, that ye which have followed me, in the regeneration **when the Son of man shall sit in the throne of his glory, ye also shall sit upon twelve thrones, judging the twelve tribes of Israel**. (Matthew 19:28, KJV)*

And please remember, as we have already read, the kingdom of heaven talked about, is more than just our earth and atmosphere, so the ==glorious rulership== extends beyond just earth, for none but Christ had been there, and it includes the sun and stars in the discussion at the mysterious change of man, in 1Cor15. For we must be changed to enter it (from flesh to spirit existence).

> *There are also **heavenly bodies** and there are **earthly bodies**; but the splendor of the heavenly bodies is one kind, and the splendor of the earthly bodies is another. **The sun** has one kind of splendor, **the moon** another and **the stars another**; and star differs from star in splendor.*
> *So will it be with the resurrection of the dead. The body that is sown is perishable, it is raised imperishable; it is sown in dishonor, ==it is raised in glory==; it is sown in weakness, it is raised in power; it is sown a natural body, it is raised a **spiritual body.***
> *If there is a natural body, there is also a spiritual body. So it is written: "The first man Adam became a living being"; the last Adam, a life-giving spirit. The spiritual did not come first, but the natural, and after that the spiritual. The **first man was of the dust of the earth, the second man from heaven**.*
> *(1 Corinthians 15:40–47, NIV)*

Man has now touched that heaven by travelling to the moon—a phenomenal accomplishment. *(Genesis 11:5–7: "And nothings shall be impossible for them")*. And when Christ spoke on earth: *John 3:13 (NIV):* "==**No one has ever gone into heaven**== except the one who came from heaven—the Son of Man."

> ***Wisdom From the Spirit***
> *We do, however, speak a message of wisdom among the mature, but not the wisdom of this age or of the rulers of this age, who are coming to nothing. No, we speak of God's secret wisdom, a wisdom that has been hidden* **and that God destined ==for our glory== before time**

began. None of the rulers of this age understood it, for if they had, they would not have crucified the Lord of glory. However, as it is written:
==*"No eye has seen,*==
==*no ear has heard,*==
==*no mind has conceived*==
==*what God has prepared for those who love him"*==—
but God has revealed it to us by his Spirit.
The Spirit searches all things, even the deep things of God. (1 Corinthians 2:6–10, NIV)

What a revelation! What an inheritance, we thank you God. I can't wait for the time when all the billions of people who are not yet Christians realize they are part of this too.

God will give it to them once they join in—perhaps **at the last hour of the day and earn the same pay. <u>I love telling them this</u>** and cringe when I hear us warn them they are going to roast in hell for eternity if they don't accept Jesus the way we tell them too. **Jesus knows the best time to tell them, and tell them he will!** And join in they will.

What a monster we make God out to be. A friend of mine named George once told me an analogy about this. It is something worth all of us considering. Suppose a man bought a very expensive sheep dog. But the dog was a great disappointment, and dumber than a sac full of hammers. He tried for weeks to get the dog to be what it should be, but it just failed. Now would this shepherd be justified in taking that dog, tying it to a spit for punishment, and slowly roasting it over a fire? Not too close, not hot enough to kill it, just close enough to absolutely torment it in pain. And suppose he did this for days!

What an uproar that would come from us, and we would lock the shepherd up right. Yet we make our loving God to be worse, because God is supposedly going to roast unbelievers and unrepentant sinners for eternity! God forgive our foolish blindness in saying that to unbelievers. Thankfully God wills that all would come to salvation (some say because that is Gods will, then everyone will be saved. I hope so.

> *For this is good and acceptable in the sight of God our Saviour;*
> ***Who will have all men to be saved****, and to come unto the knowledge of the truth.*
> *For there is one God, and one mediator between God and men, the man Christ Jesus;*
> ***Who gave himself a ransom for all, to be testified in due time.*** *(1 Timothy 2:3–6, KJV)*[11]

And here is the rebellion of the original light-bringer (the original now failed officer of the morning star), followed by another example of God flipping back between the present and the past (Satan's rebellion and man following Satan's footsteps).

> *How you have fallen from heaven,*
> *O morning star, son of the dawn!*
> *You have been cast down to the earth,*
> *you who once laid low the nations!*
> ==*You said in your heart,*==
> ==*"I will ascend to heaven;*==
> ==*I will raise my throne*==
> ==*above the stars of God;*==
> *I will sit enthroned on the mount of assembly,*
> *on the utmost heights of the sacred mountain.*
> *I will ascend above the tops of the clouds;*
> ==*I will make myself like the Most High."*==
> *But you are brought down to the grave,*
> *to the depths of the pit. (Isaiah 14:12–15, NIV)*

Below, God compares the king of Tyre to Satan (for the physical king of Tyre was not in the garden of Eden, nor was he cast down out of heaven like Satan was).

> *The word of the LORD came to me: [12] "Son of man, take up a lament concerning the king of Tyre and say to him: 'This is what the Sovereign LORD says:*

"'You were the model of perfection,
full of wisdom and perfect in beauty.
You were in Eden,
the garden of God;
every precious stone adorned you:
ruby, topaz and emerald,
chrysolite, onyx and jasper,
sapphire, turquoise and beryl.
Your settings and mountings were made of gold;
on the day you were created they were prepared.
You were anointed as a guardian cherub,
for so I ordained you.
You were on the holy mount of God;
you walked among the fiery stones.
<u>You were blameless in your ways</u>
<u>from the day you were created</u>
<u>till wickedness was found in you.</u>
<u>Through your widespread trade</u>
<u>you were filled with violence,</u>
<u>and you sinned.</u>
<u>So I drove you in disgrace from the mount of God,</u>
<u>and I expelled you, O guardian cherub</u>,
from among the fiery stones.
<u>Your heart became proud</u>
<u>on account of your beauty,</u>
<u>and you corrupted your wisdom</u>
<u>because of your splendor.</u>
<u>So I threw you to the earth;</u>
I made a spectacle of you before kings.
By your many sins and dishonest trade
you have desecrated your sanctuaries.
So I made a fire come out from you,
and it consumed you,
and I reduced you to ashes on the ground
in the sight of all who were watching.
(Ezekiel 28:11–18, NIV)

I think God uses these analogies to show us similarities to the past where Satan and his ways are corrupt and selfish. Man (like the actual king of Tyre) would follow in those footsteps and echo the ways of Satan, which preceded our existence (before what I think is the re-creation of Genesis 1; see later explanation).

The kingdoms of mankind (see Daniel's image of the golden-headed statue), except for Christ, follow the path of the prince of the power of the air (Satan) and rule in the same boastful, selfish evil way.

A description of the succession of ruling kingdoms on earth, as described in the book of Daniel, most probably representing the Babylonian (gold), Medo-Persian (silver), Greek-Macedonian (bronze), and split Roman (iron and clay).

> *"You looked, O king, and there before you stood a large statue—an enormous, dazzling statue, awesome in appearance. The head of the statue was made of pure gold, its chest and arms of silver, its belly and thighs of bronze, its legs of iron, its feet partly of iron and partly of baked clay. While you were watching, a rock was cut out, but not by human hands. It struck the statue on its feet of iron and clay and smashed them. Then the iron, the clay, the bronze, the silver and the gold were broken to pieces at the same time and became like chaff on a threshing floor in the summer. The wind swept them away without leaving a trace. But the rock that struck the statue became a huge mountain and filled the whole earth. (Daniel 2:31–35, NIV)*

Pride and loftiness are the downfall of man and fallen spirit beings, but **God is High and worthy of worship and all praise!**

The rock of Christ will replace the image and replace the kingdoms of man.

> *"After you, another kingdom will rise, inferior to yours. Next, a third kingdom, one of bronze, will rule over*

the whole earth. Finally, there will be a fourth kingdom, strong as iron—for iron breaks and smashes everything—and as iron breaks things to pieces, so it will crush and break all the others. Just as you saw that the feet and toes were partly of baked clay and partly of iron, so this will be a divided kingdom; yet it will have some of the strength of iron in it, even as you saw iron mixed with clay. As the toes were partly iron and partly clay, so this kingdom will be partly strong and partly brittle. And just as you saw the iron mixed with baked clay, so the people will be a mixture and will not remain united, any more than iron mixes with clay.

"In the time of those kings, the God of heaven will set up a kingdom that will never be destroyed, nor will it be left to another people. It will crush all those kingdoms and bring them to an end, but it will itself endure forever. *This is the meaning of the vision of the rock cut out of a mountain, but not by human hands—a rock that broke the iron, the bronze, the clay, the silver and the gold to pieces.*

"The great God has shown the king what will take place in the future. ==The dream is true and the interpretation is trustworthy==*." (Daniel 2:39–45, NIV)*

In the flashbacks of Isaiah 14 and Ezekiel 28, I understand that Satan, the Antichrist, is the one referred to in these stories (original in the role of self-importance and rebellion) as the prideful being that attempted to raise himself up above the true God and is now the collapsed but still powerful "prince of darkness and so-called power of the air."

God is showing us how we naturally follow the pattern of Satan and reject God. We need to be changed, and when God sees we choose to be like Him, he will wash us in the blood of the Lamb, and we will be worthy to take of the tree of life, and live forever as now spirit essence like God.

Wherein in time past ye walked according to the course of this world, **according to the prince of the power of the air, the spirit that now worketh in the children of disobedience***:*
Among whom also we all had our conversation in times past in the lusts of our flesh, fulfilling the desires of the flesh and of the mind; and were by nature the children of wrath, even as others.
But God, who is rich in mercy, for his great love wherewith he loved us,
Even when we were dead in sins, **hath quickened us together with Christ (by grace ye are saved).** *(Ephesians 2:2–5, KJV)*

And do this, understanding the present time. The hour has come for you to wake up from your slumber, because our salvation is nearer now than when we first believed. The night is nearly over; the day is almost here. So **let us put aside the deeds of darkness and put on the armor of light.** *Let us behave decently, as in the daytime, not in orgies and drunkenness, not in sexual immorality and debauchery, not in dissension and jealousy. Rather, clothe yourselves with the Lord Jesus Christ, and do not think about how to gratify the desires of the sinful nature. (Romans 13:11–14, NIV)*

This then is the message which we have heard of him, and declare unto you, that **God is light**, *and in him is no darkness at all. (1 John 1:5, KJV)*

For Christ is now our true morning star and true light-bringer that replaced the rebellious one.

> ## Why is Jesus Called the "Morning Star"?
> Randy Alcorn
>
> 🏠 › Is Jesus God? › Names of Jesus › Why is Jesus Called the "Morning Star"?
>
> I think it relates to HOPE and his immanent second coming. When Venus rises, it means the sun will follow very soon (in a morning, usually within an hour or two, sometimes just a matter of minutes). Christ's coming—and in a broad sense this could apply to both his first and his second coming—means/will mean God's light is about to shine forever on the universe, making all wrongs right, wiping away all tears, and fulfilling Revelation 21, with the creation of the new heavens and new earth, etc. On a long dark night, the appearance of the morning star means daybreak is imminent. In the long dark night of suffering on earth, Jesus being seen as the morning star means the eternal morning is about to dawn. Hence, Christ as the morning star is a picture of great promise and hope.
>
> About Randy Alcorn
>
> In one of my books I call Jesus the Morning Star and several readers have gotten very upset, wondering why I would give Jesus a name that belongs to Satan. They are of course referring to Isaiah 14:12: "How you have fallen from heaven, O morning star, son of the dawn! You have been cast down to the earth, you who once laid low the nations!"
>
> The point is that Morning Star was a name for Lucifer before his fall, and there is no inconsistency with two very different beings called by the same name. Lucifer was a creature of beauty and power. Christ is God, the Creator, beautiful and powerful beyond measure, the one Lucifer rebelled against. But the name Morning Star is not tainted—it is Satan who is tainted. Obviously this is the case, or Morning Star wouldn't be used of Christ as it is in Revelation 22:1-15, nor used in a positive way as it is in 1 Peter 1 and Revelation 2.
>
> *Taken from "Jesus, Venus, and the Morning Star" by Randy Alcorn, Eternal Perspective Ministries, 39085 Pioneer Blvd., Suite 206, Sandy, OR 97055, 503-668-5200, www.epm.org*

Figure 13. Jesus role as, Star of the Morning (Jesus, Venus, and the Morning Star", by Randy Alcorn, Eternal Perspective Ministries.)

Comment on Revelation 22:16: Christ, the descendant of David, is the leader of the morning stars, and it is not the planet Venus but the true stars that bring light.

Jesus is the leader of these Morning Stars, as we share in bringing that light to all who will hear.

SPACE THE TRUE FRONTIER!

Figure 14. Jesus Son of the Morning Star- Removes Enemies (The IVP Bible Background Commentary Old and New Testaments)

> **The IVP Bible Background Commentary: Old and New Testaments**　　　　22:15
>
> 22:16. "Root of David" comes from the "stem of Jesse" (David's father) in Isa 11:1 - the shoot that would spring up from the stump of David's lineage, after his descendants had lost the throne. Some commentators suggest that "root" reverses the image, making him David's source. The morning star is Venus, herald of the dawn (cf. Rev 2:28), which in this case probably alludes to Nu 24:17, the star descended from Jacob (Israel) and destined to reign and crush the enemies of God's people. (The *Dead Sea Scrolls also applied Nu 24:17 to a conquering *Messiah.)

Again, *Revelation 22:16 (NIV):* "*I, Jesus, have sent my angel to give you this testimony for the churches.* **I am the Root and the Offspring of David, and the bright Morning Star**."

Jesus, who was in times past the Word God, is that true light of the morning—the light of men:

> ### The Word Became Flesh
> *In the beginning was the Word, and the Word was with God, and the Word was God. He was with God in the beginning.*
> *Through him all things were made; without him nothing was made that has been made. In him was life, and* **that life was the light of men.** *The light shines in the darkness, but the darkness has not understood it.*
> *There came a man who was sent from God; his name was John. He came as a witness to testify* **concerning that light**, *so that through him all men might believe. He himself was not the light; he came only as a witness to the light.* **The true light that gives light to every man** *was coming into the world.*
> *He was in the world, and though the world was made through him, the world did not recognize him. He came to that which was his own, but his own did not receive him. Yet to all who received him, to those who believed in his name, he gave the right to become children of*

> *God—children born not of natural descent, nor of human decision or a husband's will, but born of God. The Word became flesh and made his dwelling among us. We have seen his glory, the glory of the One and Only, who came from the Father, full of grace and truth. (John 1:1–14, NIV)*

As already stated, Satan, before his rebellion, was given a very high position (morning star or light-bringer), which was taken from him. Yet he still believes that Christ and everyone else should worship him as God, and I believe he is trying to convince everyone else to rebel like he did. *"He will oppose and will exalt himself over everything that is called God or is worshiped, so that* **he sets himself up in God's temple, proclaiming himself to be God"** *(2 Thessalonians 2:4, NIV).* Christ refused to worship him, but much of mankind may unknowingly be doing that very thing.

> *Again, the devil took him to a very high mountain and showed him all the kingdoms of the world and their splendor. "All this I will give you," he said,* ***"if you will bow down and worship me."*** *Jesus said to him, "Away from me, Satan! For it is written:* **'Worship the Lord your God, and serve him only.'"** *(Matthew 4:8–10, NIV)*

Satan apparently (speculation on my part) could not stand the idea that mere humans could eventually inherit an eternal spiritual existence, as part of the very likeness of God, in God's very family—adopted sons of God (and therefore above the angels including the fallen archangel Satan).

At one in a spiritual unity with Christ and the Father, the everlasting family of God, the temple of God! Ruling with God in space (new heavens, the true Israel).

In the ultimate temple of God, there will be God the Father, Jesus the Christ, and His Bride (the church).

*Then I saw a new heaven and a new earth, for the first heaven and the first earth had passed away, and there was no longer any sea. I saw the Holy City, **the new Jerusalem, coming down out of heaven from God, prepared as a bride** beautifully dressed for her husband. And I heard a loud voice from the throne saying, "Now the dwelling of God is with men, and he will live with them. They will be his people, and God himself will be with them and be their God. He will wipe every tear from their eyes. There will be no more death or mourning or crying or pain, for the old order of things has passed away." He who was seated on the throne said, "I am making everything new!" Then he said, "**Write this down, for these words are trustworthy and true**." (Revelation 21:1–5, NIV)*

*And I saw no temple therein: **for the Lord God Almighty and the Lamb are the temple** of it. And the city had no need of the sun, neither of the moon, to shine in it: for the **glory of God did lighten it,** and **the Lamb is the light thereof.** (Revelation 21:22–23, KJV)*

*Don't you know that **you yourselves are God's temple and that God's Spirit lives in you?** If anyone destroys God's temple, God will destroy him; for God's temple is sacred, and you are that temple. (1 Corinthians 3:16–17, NIV)*

The book of John chapter 17 reveals our future oneness, as we truly become like God in His eternal family.

And now I am no more in the world, but these are in the world, and I come to thee. Holy Father, keep through thine own name those whom thou hast given me, that they may be one, as we are. While I was with them in the world, I kept them in thy name: those that thou gav-

est me I have kept, and none of them is lost, but the son of perdition; that the scripture might be fulfilled.

And now come I to thee; and these things I speak in the world, that they might have my joy fulfilled in themselves. I have given them thy word; and the world hath hated them, because they are not of the world, even as I am not of the world. I pray not that thou shouldest take them out of the world, but that thou shouldest keep them from the evil. They are not of the world, even as I am not of the world. Sanctify them through thy truth: thy word is truth. As thou hast sent me into the world, even so have I also sent them into the world. And for their sakes I sanctify myself, that they also might be sanctified through the truth. ==Neither pray I for these alone, but for them also which shall believe on me through their word;== **==That they all may be one; as thou, Father, art in me, and I in thee, that they also may be one in us==**: *that the world may believe that thou hast sent me. And the glory which thou gavest me I have given them; that they may be one, even as we are one: I in them, and thou in me, that they may be made perfect in one; and that the world may know that thou hast sent me, and hast loved them, as thou hast loved me. Father, I will that they also, whom thou hast given me, be with me where I am; that they may behold my glory, which thou hast given me: for thou lovedst me before the foundation of the world. O righteous Father, the world hath not known thee: but I have known thee, and these have known that thou hast sent me.* ==And I have declared unto them thy name, and will declare it: that the love wherewith thou hast loved me may be in them, and I in them==. *(John 17:11–26, KJV)*

Angels do not have that promise (only men and women, Genesis 1:27), so perhaps that is why archangel rebelled and decided he was god?

It is my thinking (speculative on this) that the six days of the Genesis creation was a re-creation of the previously destroyed earth and universe. Physical human mankind was new and made to eventually be changed to spiritual like God. Apparently, it was in a state of decay (tohu and wabohu)[18] like the rest of the universe, but God fixed earth to make it habitable for man, and the seeds of His plan to make a new universe with His Family.

> *In the beginning of God's preparing the heavens and the earth—the earth hath existed waste and void, and darkness is on the face of the deep, and the Spirit of God fluttering on the face of the waters (Genesis 1:1–2, YLT)*

> *I looked at the earth,*
> *and **it was formless and empty;***
> *and at the heavens,*
> *and **their light was gone**. (Jeremiah 4:23, NIV)*

> *I looked at the earth—*
> ***it was back to pre-Genesis chaos and emptiness.***
> *I looked at the skies,*
> *and not a star to be seen. (Jeremiah 4:23, MSG)*

Figure 15. Light had to be restored (ISA)

Is this scripture (Jeremiah 4:23) talking about a light in the heavens that no longer was there? I think so (needing the re-creation, or restoration, to past light, pre-Genesis).

God started out in Genesis by creating light again.

> *Then God said,* **"Let us make man in our image, in our likeness**, *and let them rule over the fish of the sea and the birds of the air, over the livestock, over all the earth, and over all the creatures that move along the ground."*
> *So God created man in his own image,*
> *in the image of God he created him;*
> *male and female he created them. (Genesis 1:26–27, NIV)*

It is important to note that everything else created on earth, was created after its 'kind', and not just a likeness. Where mankind is not after the godkind but after the likeness (still amazing and **embarrassingly almost blasphemous**).

God has promised us so much, how wonderful, and barely comprehensible.

> *"I said, 'You are "gods";*
> *you are all sons of the Most High.' (Psalms 82:6, NIV)*

> *Jesus answered them, "Is it not written in your Law, 'I have said you are gods'? If he called them 'gods,' to whom the word of God came—and the Scripture cannot be broken—what about the one whom the Father set apart as his very own and sent into the world? Why then do you accuse me of blasphemy because I said, 'I am God's Son'? (John 10:34–36, NIV)*

The Pharisees did not see that mankind is in the image of God, eventually to be part of His spiritual self (literally adopted sons of God, his Family).

From God's word, worth repeating:

> *I consider that our present sufferings are not worth comparing with the glory that will be revealed in us.*

SPACE THE TRUE FRONTIER!

The creation waits in eager expectation for the sons of God to be revealed. For the creation was subjected to frustration, not by its own choice, but by the will of the one who subjected it, in hope that the creation itself will be liberated from its bondage to decay and brought into the glorious freedom of the children of God. (Romans 8:18–21)

God is a spiritual oneness now composed of Father and the only begotten Son that eventually combines with the bride (church).

The Word of Life
That which was from the beginning, which we have heard, which we have seen with our eyes, which we have looked at and our hands have touched—this we proclaim concerning the Word of life. The life appeared; we have seen it and testify to it, and we proclaim to you the eternal life, which was with the Father and has appeared to us. We proclaim to you what we have seen and heard, so that you also may have fellowship with us. ***And our fellowship is with the Father and with his Son, Jesus Christ. [4] We write this to make our joy complete****. (1 John 1:1–3, NIV)*

That they all may be one; as thou, Father, art in me, and I in thee, that they also may be one in us. John 17; [21

Ultimately, the **Spirit and the bride go forth** and expand the spiritual kingdom in a new glorious universe. (The Spirit is the essence of Jesus and the Father that is in us). Romans 8:9–11 The New World Translation

> **9** However, you are in harmony, not with the flesh, but with the spirit,ᵐ **if God's spirit truly dwells in you**. But if anyone does not have Christ's spirit, this person does not belong to him. **10** But **if Christ is in union with you**,ⁿ the body is dead because of sin, but the spirit is life because of righteousness. **11** **If, now, the spirit of him who raised up Jesus from the dead dwells in you**, the one who raised up Christ Jesus from the dead° will also make your mortal bodiesᵖ alive **through his spirit that resides in you**.

So it is Jesus, the Father, and the bride, (all existing as a spiritual family) that go forth to create new fruit by passing that Spirit on (Living Water for the thirsty).

Again, *Revelation 21:1–4 (NIV)*

> *Then I **saw a new heaven and a new earth**, for the first heaven and the first earth had passed away, and there was no longer any sea. I saw the Holy City, the new Jerusalem, coming down out of heaven from God, prepared as a bride beautifully dressed for her husband. And I heard a loud voice from the throne saying, "Now the dwelling of God is with men, and he will live with them. They will be his people, and God himself will be with them and be their God. He will wipe every tear from their eyes. There will be no more death or mourning or crying or pain, for the old order of things has passed away."*
>
> *The Spirit and the bride say, "Come!" And let him who hears say, "Come!" Whoever is thirsty, let him come; and whoever wishes, let him take the free gift of the water of life.* (Revelation 22:17, NIV)

My point is that God is sharing his existence, with adopted sons that he is bringing into his spiritual family. We will live together in a spiritual oneness (atonement) in a new and perfect universe.

There has always been great opposition to God's plan, especially from the fallen archangel now Satan the devil. Satan walks about heaven making himself out to be god, but he is not. He tried to get Christ to worship him, proving he is god, but that did not happen!

> *[Then Jesus was led by the Spirit into the desert to be tempted by the devil. After fasting forty days and forty nights, he was hungry. The tempter came to him and said, "If you are the Son of God, tell these stones to become bread."*
> *Jesus answered, "It is written: 'Man does not live on bread alone, but on every word that comes from the mouth of God.'"*
> *Then the devil took him to the holy city and had him stand on the highest point of the temple. "If you are the Son of God," he said, "throw yourself down. For it is written:*
>
> *"'He will command his angels concerning you,*
> *and they will lift you up in their hands,*
> *so that you will not strike your foot against a stone.'"*
>
> *Jesus answered him, "It is also written: 'Do not put the Lord your God to the test.'"*
> *Again, the devil took him to a very high mountain and showed him* **all the kingdoms of the world and their splendor. "All this I will give you,"** *he said,* **"if you will bow down and worship me."**
> *Jesus said to him, "Away from me, Satan!* **For it is written: 'Worship the Lord your God, and serve him only.'"**
> *Then the devil left him, and angels came and attended him. (Matthew 4:1–11, NIV, cf Mark 1:12, 13; Luke 4:1–13)*

And so the man Jesus overcame Satan! Something Satan was probably sure could never happen!

Satan hasn't given up. He desires our worship, and he still today interjects himself into the temple of God, demanding such. God allows it because it works toward our advancement to realize we must also reject this mighty prince of the power of the air. **Just as our elder brother said, "Get behind me Satan, and worship the Lord only," so must we.**

A major part of writing this book is to warn us about what I think is the greatest deception in all history—**the agenda to displace the sons of God with the false god, Satan.** I am talking about something that, to many, appears so right but is so wrong. That is a Satan-inspired manmade agenda to make out the Godhead as a Trinity, excluding the bride. It is my primary hope that once the ancient Trinity deception is revealed, we can all see our amazing wonderful human potential, that of being one in the spirit in the likeness of God, to be part of the very family of God—Father-Son-Bride setting about dressing the garden (universe).

The current view of Christianity in general is a Trinity Godhead with, Father, Son, and Spirit persons instead of bride. To support this evil (Satan-inspired) agenda, man has changed the scriptures, injected, and added words, and knowingly improperly translated words.

And like a hammer of injustice, the errant churches have implemented a restriction for new potential members that they must accept this Trinity doctrine or be rejected from the family of God (church). This is found in almost every statement of beliefs.

Yet Christ never used, or ever heard, the word *Trinity*. It is nowhere to be found in the word of God—the Bible!

> *I hope you will put up with a little of my foolishness; but you are already doing that. I am jealous for you with a godly jealousy.* ***I promised you to one husband, to Christ****, so that I might present you as a pure virgin to him.* ***But I am afraid that just as Eve was deceived by the serpent's cunning, your minds may somehow be led astray from your sincere and pure devotion to Christ.*** *For if someone comes to you and preaches a*

Jesus other than the Jesus we preached, or if you receive a different spirit from the one you received, or a different gospel from the one you accepted, you put up with it easily enough.
For such men are false apostles, deceitful workmen, masquerading as apostles of Christ. And no wonder, **for Satan himself masquerades as an angel of light.** *It is not surprising, then, if his servants masquerade as servants of righteousness. Their end will be what their actions deserve. 2 Corinthians 11:1–4, 13–15 (NIV)*

I will attempt to show the false god, or Antichrist, or Satan the devil, has been doing everything he can to hide from us this destiny. He makes himself out to be god, seeking our worship. He displaces us from the unity with God that God has promised. Let me begin.

CHAPTER 1
THE PARABLE OF THE SOWER

If God truly made space for us and intends to be at one with us, why didn't he tell us? I think he did, but for a time, *we just can't know.*

This chapter shows the importance of faith, of seeing, and of hearing. It is a story of how God will reveal this "**mystery**" in His good time. Before that, we are protected by blindness, so we can be forgiven.

Romans 16:25 (NIV) Now to him who is able to establish you by my gospel and the proclamation of Jesus Christ, according to the revelation of the mystery hidden for long ages past, [26] but now revealed and made known through the prophetic writings by the command of the eternal God, so that all nations might believe and obey him— [27] to the only wise God be glory forever through Jesus Christ! Amen.

The next few verses are, in my opinion an example of this point and explain how when we see space, we don't grasp it is our future home and how although we all grew up in families, we still don't really understand the significance of becoming adopted into God's very family.

> *At daybreak the council of the elders of the people, both the chief priests and teachers of the law, met together, and Jesus was led before them. "If you are the Christ," they said, "tell us."*
> *Jesus answered, "If I tell you, you will not believe me, and if I asked you, you would not answer. But from now on, the Son of Man will be seated at the right hand of the mighty God."*
> *They all asked, "Are you then the Son of God?"*
> *He replied, "You are right in saying I am."*

> *Then they said, "Why do we need any more testimony? We have heard it from his own lips." (Luke 22:66–71, NIV)*

I hope that most of the Pharisees did not believe Him, but if some did, they ==would not reveal it==, and they would be ==in grave danger of a condemning judgement.==

> *He replied, "I will also ask you a question. Tell me, John's baptism—was it from heaven, or from men?" They discussed it among themselves and said, "If we say, 'From heaven,' he will ask, 'Why didn't you believe him?' But if we say, 'From men,' all the people will stone us, because they are persuaded that John was a prophet."*
> *So they answered, =="We don't know where it was from."==*
> *Jesus said, "Neither will I tell you by what authority I am doing these things." (Luke 20:3–8, NIV)*

Yet a man born blind, can have his eyes opened very quickly. Just like you can have yours opened, and as I hope mine have been opened.

If you have, then, and only then, can you see the mystery of the Kingdom and know it is true, in your heart.

From John 9, the man below had been deliberately blind from his birth to reveal how much <u>we need God before we will ever see anything</u>!

> *Jesus heard that they had thrown him out, and when he found him, he said, "Do you believe in the Son of Man?"*
> *"Who is he, sir?" the man asked. "Tell me so that I may believe in him."*
> *Jesus said, "You have now <u>seen</u> him; in fact, he is the one <u>speaking with you</u>."*
> *Then the man said, "Lord, I believe," and he worshiped him.*

> *Jesus said, "For judgment I have come into this world, so that the blind will see <u>and those who see will become blind.</u>"*
> *Some Pharisees who were with him heard him say this and asked, "What? Are we blind too?"*
> *Jesus said, <mark>"If you were blind, you would not be guilty of sin</mark>; but now that you claim you can see, your guilt remains. (John 9:35–41, NIV)*

When we look to the stars now in the twenty-first century, are we seeing the physical manifestation of the "glory of Christ in heaven" awaiting us?

Only in the last few centuries of the Christian era has it been possible to see space with much understanding. May we grow in grace and knowledge and see it ever more clearly as the day approaches. I think science has been an instrument of God in bringing us this additional sight into the former glory of Christ in space (like the Hubble Space Telescope reveals God's glory).

God has repeatedly explained to us that He is expanding His family to include many adopted sons. But we don't seem to fathom the significance of that. It is better than finding out we have a distant relative who has an inheritance of a billion dollars waiting for us that we didn't yet know about. This is much more—it is eternal sharing in the glory of Christ throughout the universe in the very family of God!

I think God tells us of our mysterious inheritance in the previously mostly unseen parable of the sower.

> *"Listen then to what the parable of the sower means: When anyone hears the message about the kingdom* and does not understand it, **the evil one comes and snatches away what was sown in his heart**. *This is the seed sown along the path." (Matthew 13:18–19, NIV)*

> *But blessed are your eyes because they see, and your ears because they hear. For I tell you the truth, <u>many prophets</u> and <u>righteous men</u> longed to see what you see*

but did not see it, and to hear what you hear but did not hear it. (Matthew 13:16–17, NIV)

For a time, most people, including the even <u>righteous prophets</u>, have the truth about the kingdom of God hidden from them!

*<u>Ye did not choose me, but I chose you</u>, and appointed you, **that ye should go and bear fruit**, and that your fruit should abide: that whatsoever ye shall ask of the Father in my name, he may give it you. (John 15:16, ASV)*

*And he that was sown upon the good ground, this is he that **heareth** the word, **and understandeth** it; who **verily beareth fruit**, and bringeth forth, **some a hundredfold**, some sixty, some thirty. (Matthew 13:23, ASV)*

*He who belongs to God hears what God says. **The reason you do not hear is that you do not belong to God**. (John 8:47, NIV)*

*The Jews gathered around him, saying, "How long will you keep us in suspense? If you are the Christ, tell us plainly." Jesus answered, "**<u>I did tell you, but you do not believe</u>**. The miracles I do in my Father's name speak for me, [26] but you do not believe because you are not my sheep. **My sheep listen to my voice; I know them, and they follow me. I give them eternal life, and they shall never perish**; no one can snatch them out of my hand. My Father, who has given them to me, is greater than all; no one can snatch them out of my Father's hand. I and the Father are one." (John 10:24–30, NIV)*

God knows the best time to open each of our eyes and when to open our ears. Before He does, we can look right at it and not see it. We can be right next to the speaker but not hear a word. We are born blind, until the sower, sows His seed, into good soil and

adds living water, to make it grow. (more on John 9 a man born blind, until Jesus took soil, and His spittle)

On gaining sight: *"Having said this,* **he spit on the ground, made some mud with the saliva, and put it on the man's eyes"** *(John 9:6, NIV).*

On hearing: *"There some people brought to him a man who was deaf and could hardly talk, and they begged him to place his hand on the man. After he took him aside, away from the crowd, Jesus put his fingers into the man's ears.* **Then he spit and touched the man's tongue.** *He looked up to heaven and with a deep sigh said to him, "Ephphatha!" (which means, "Be opened!"). At this, the man's ears were opened, his tongue was loosened and he began to speak plainly" (Mark 7:32–35, NIV).*

I always found these actions of Christ to be strange. Why spit and why soil (mud)? <u>For me, this is the parable of the sower in action.</u> And when you realize it, I think we find God opening our eyes to His earthly and heavenly kingdom! Do you see it? Do you hear it?

Again on seeing and hearing:

> *Jesus heard that they had thrown him out, and when he found him, he said, "Do you believe in the Son of Man?" "Who is he, sir?" the man asked. "Tell me so that I may believe in him."*
> *Jesus said, "***You have now seen him***; in fact, he is the one **speaking with you**."*
> *Then the man said, "Lord, I believe,"* ==**and he worshiped him**==.
> *Jesus said, "For judgment I have come into this world, so that the blind will see and those who see will become blind." Some Pharisees who were with him heard him say this and asked, "What? Are we blind too?"*
> *Jesus said, "If you were blind, you would not be guilty of sin; but now that you claim you can see, your guilt remains. (John 9:35–41, NIV)*

It is **our righteous duty to believe,** as it was Abrahams righteousness. And even that is a gift of God. *James 2:23 (NIV): "And the* *scripture was fulfilled that says, '**Abraham believed God, and it was credited to him as righteousness**,' and he was called God's friend."*

And by the way, please notice that the man who gained his sight <u>worshipped Christ on earth</u>. Scripture does not stand for anyone but God being worshipped. Christ was God on earth as a man!

> Then the angel said to me, "Write: 'Blessed are those who are invited to the wedding supper of the Lamb!'" And he added, "These are the true words of God." **At this I fell at his feet to worship him. But he said to me, "Do not do it! I am a fellow servant with you and with your brothers who hold to the testimony of Jesus. <u>Worship God!</u>** For the testimony of Jesus is the spirit of prophecy.") (Revelation 19:9–10, NIV)

> *"Though I have been speaking figuratively,* **a time is coming when I will no longer use this kind of language but will tell you plainly about my Father**. *In that day you will ask in my name. I am not saying that I will ask the Father on your behalf. No, the Father himself loves you because you have loved me and have believed that I came from God. I came from the Father and entered the world; now I am leaving the world and going back to the Father."*
> *Then Jesus' disciples said,* **"Now you are speaking clearly and without figures of speech. Now we can see** *that you know all things and that you do not even need to have anyone ask you questions. This makes us believe that you came from God."*
> <u>**"You believe at last!"**</u> *Jesus answered. "But a time is coming, and has come, when you will be scattered, each to his own home. You will leave me all alone. Yet I am not alone, for my Father is with me.*
> *"I have told you these things, so that in me you may have peace. In this world you will have trouble. But take heart! I have overcome the world." (John 16:25–33, NIV)*

Back to what we now can see of the glory of Christ in these last days. Everyone has looked up at the night sky in amazement, especially those of us in these times, using the added knowledge from science that helps us understand the size and time of the universe right down to the micro world of quantum mechanics (where gravity waves may exist at something like 10^{-21} meters) and back out to what science estimates may be 13.75 billion light-years of expanse in the cosmos.[19]

But <u>sow</u> what? And why isn't this obvious?

> *When I came to you, brothers, I did not come with eloquence or superior wisdom as I proclaimed to you the testimony about God. For I resolved to know nothing while I was with you except Jesus Christ and him crucified. I came to you in weakness and fear, and with much trembling. My message and my preaching were not with wise and persuasive words,* ==**but with a demonstration of the Spirit's power**==*, so that your* ==**faith might not rest on men's wisdom, but on God's power.**==
>
> *Wisdom From the Spirit*
> *We do, however, speak a message of wisdom among the mature, but not the wisdom of this age or of the rulers of this age, who are coming to nothing. No, we speak of God's secret wisdom,* ==**a wisdom that has been hidden and that God destined for our glory before time began**==*. None of the rulers of this age understood it, for if they had, they would not have crucified the Lord of glory. However, as it is written:*
> <mark style="background-color: lightblue">*"No eye has seen,*</mark>
> *no ear has heard,*
> *no mind has conceived*
> <mark style="background-color: lightblue">*what God has prepared for those who love him"*</mark>—
> *but God has revealed it to us by his Spirit.*
> *The Spirit searches all things, even the deep things of God. (1 Corinthians 2:1–10, NIV)*

The mystery about what God is doing in a nutshell, or should I say "wheat-shell"? We must have faith, and from the gift of God, we shall understand. I cannot show it to you because **no man can see it unless God chooses to open our eyes**. It is my hope, that God will show you through these words and knowledge he has given me.

Christ said, ==*"The words I have spoken to your are Spirit, and they are life!"*==

Christ says the words are spirit and life, not a person is the spirit and life. The only person here is Christ! The words and the powerful will of God are the Spirit! God in action.

Many Disciples Desert Jesus

60 On hearing it, many of his disciples* said, "This is a hard teaching. Who can accept it?"*

61 Aware that his disciples were grumbling about this, Jesus said to them, "Does this offend you?* 62 What if you see the Son of Man* ascend to where he was before!* 63 The Spirit gives life;* the flesh counts for nothing. ==The words I have spoken to you are spirit*[Or *Spirit*] and they are life.== 64 Yet there are some of you who do not believe." For Jesus had known* from the beginning which of them did not believe and who would betray him.* 65 He went on to say, "This is why I told you that no one can come to me unless the Father has enabled him."*

Gods word is Spirit (John 6:63). **And those spiritual words coming out of Jesus tells us,** *"No one can come to me unless the Father has enabled him."*

If you ask any Christian if he/she has ever heard God has prepared the universe for us, they would probably say, "That is news to me." I hope it makes sense now. I also hope we can appreciate how careful God is in picking the best possible time for us to hear it.

I find that Christianity seems to see God as trying desperately to get people to believe he exists. I say, that would be easy but not good for us. He could just fill the sky with his face and say, "Boo!" Why not? It would certainly get our attention, and maybe even scientists would be finally impressed by what he can show them.

God has protected us by waiting for our individual best time to hear the truth of his coming kingdom-family. Perhaps it is your time

SPACE THE TRUE FRONTIER!

now. It is not an easy thing when we first hear; there are times and conditions where God chooses **not** to reveal Himself!

Below is the surprising fact that God hides certain things from us. There is a time and a place for everything, and notwithstanding the great commission, it is up to God to decide the best time to reveal His mystery.

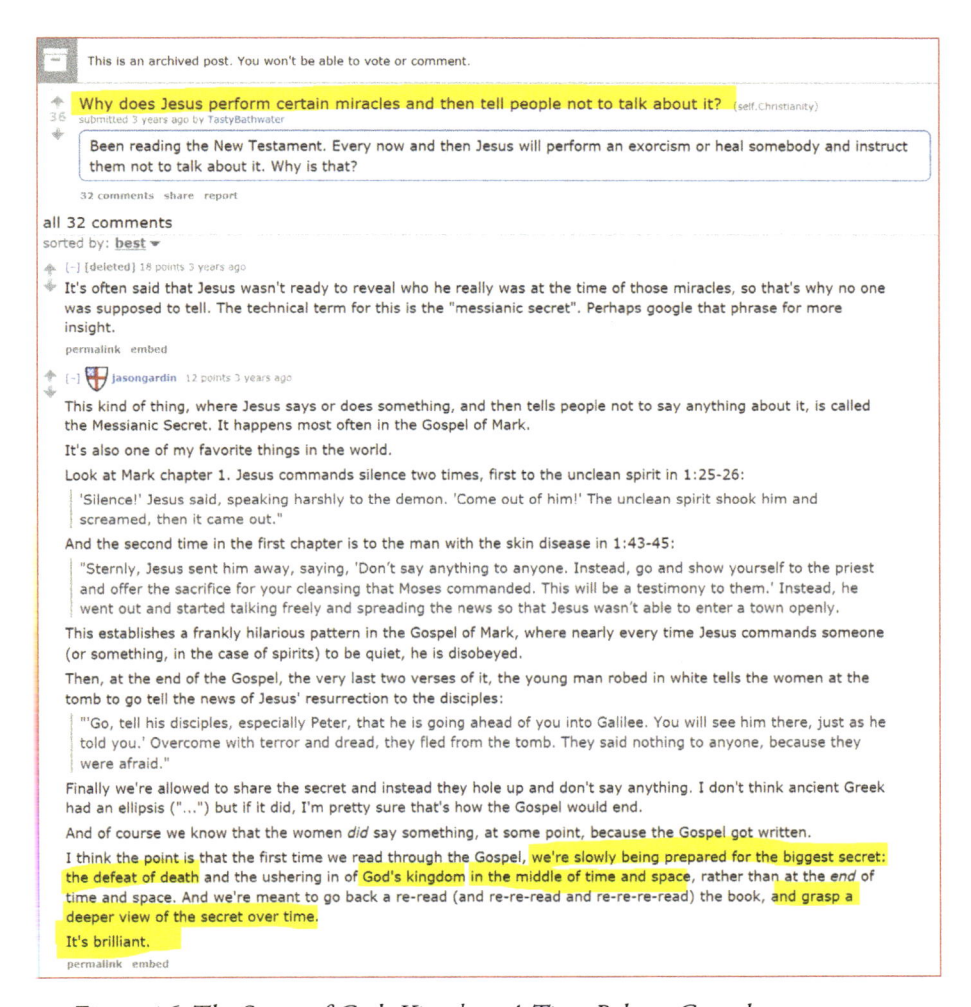

> This is an archived post. You won't be able to vote or comment.
>
> **Why does Jesus perform certain miracles and then tell people not to talk about it?** (self.Christianity)
> 36 submitted 3 years ago by TastyBathwater
>
> > Been reading the New Testament. Every now and then Jesus will perform an exorcism or heal somebody and instruct them not to talk about it. Why is that?
> >
> > 32 comments share report
>
> all 32 comments
> sorted by: best ▼
>
> [–] [deleted] 18 points 3 years ago
> It's often said that Jesus wasn't ready to reveal who he really was at the time of those miracles, so that's why no one was supposed to tell. The technical term for this is the "messianic secret". Perhaps google that phrase for more insight.
> permalink embed
>
> [–] jasongardin 12 points 3 years ago
> This kind of thing, where Jesus says or does something, and then tells people not to say anything about it, is called the Messianic Secret. It happens most often in the Gospel of Mark.
> It's also one of my favorite things in the world.
> Look at Mark chapter 1. Jesus commands silence two times, first to the unclean spirit in 1:25-26:
>
> > 'Silence!' Jesus said, speaking harshly to the demon. 'Come out of him!' The unclean spirit shook him and screamed, then it came out."
>
> And the second time in the first chapter is to the man with the skin disease in 1:43-45:
>
> > "Sternly, Jesus sent him away, saying, 'Don't say anything to anyone. Instead, go and show yourself to the priest and offer the sacrifice for your cleansing that Moses commanded. This will be a testimony to them.' Instead, he went out and started talking freely and spreading the news so that Jesus wasn't able to enter a town openly.
>
> This establishes a frankly hilarious pattern in the Gospel of Mark, where nearly every time Jesus commands someone (or something, in the case of spirits) to be quiet, he is disobeyed.
> Then, at the end of the Gospel, the very last two verses of it, the young man robed in white tells the women at the tomb to go tell the news of Jesus' resurrection to the disciples:
>
> > "'Go, tell his disciples, especially Peter, that he is going ahead of you into Galilee. You will see him there, just as he told you.' Overcome with terror and dread, they fled from the tomb. They said nothing to anyone, because they were afraid."
>
> Finally we're allowed to share the secret and instead they hole up and don't say anything. I don't think ancient Greek had an ellipsis ("...") but if it did, I'm pretty sure that's how the Gospel would end.
> And of course we know that the women *did* say something, at some point, because the Gospel got written.
> I think the point is that the first time we read through the Gospel, we're slowly being prepared for the biggest secret: the defeat of death and the ushering in of God's kingdom in the middle of time and space, rather than at the *end* of time and space. And we're meant to go back a re-read (and re-re-read and re-re-re-read) the book, and grasp a deeper view of the secret over time.
> It's brilliant.
> permalink embed

Figure 16. The Secret of Gods Kingdom-A Time Release Capsule (https://www.reddit.com/r/Christianity/comments/1jzhv5/why_does_jesus_perform_certain_miracles_1848)

"He replied, 'The knowledge of the secrets of the kingdom of heaven has been given to you, but not to them.'" (Matthew 13:11, NIV) Yes, God hides this mystery, until it is our time. I hope the Father has opened our eyes with Christ so we can now finally see what God is doing with us. Spiritually, He is growing his family, the Spirit Father, the only begotten Spirit Son (Christ), and us (the spirit bride—adopted sons and daughters). We then go forth and become more seed (fruit), and the kingdom on earth and in space increases eternally over at least 13.7 billion light-years of God stuff! Amazing! Wow.

But tremendous forces are against us understanding this parable. The very prince of the power of the air (Satan) has a plan to steal the seed before it can grow or choke the plant we are becoming to remove anything we see or hear. God will not let that happen. For blessed are your eyes, and we will be part of the harvest.

John 4:34 (NIV) [34] "My food," said Jesus, "is to do the will of him who sent me and to finish his work. [35] Do you not say, 'Four months more and then the harvest'? I tell you, open your eyes and look at the fields! They are ripe for harvest. [36] Even now the reaper draws his wages, even now he harvests the crop for eternal life, so that the sower and the reaper may be glad together. [37] Thus the saying 'One sows and another reaps' is true. [38] I sent you to reap what you have not worked for. Others have done the hard work, and you have reaped the benefits of their labor."

CHAPTER 2: NOT THREE

Christ said in John 10:30 (NIV), *"I and the Father are one."*
Satan denies these two. Note, only two personalities are denied. *"Who is the liar? It is the man who denies that Jesus is the Christ. Such a man is **the antichrist—he denies the Father and the Son**" (1 John 2:22, NIV).*

And from the early renewed, Old Testament Covenant with God:

Exodus 34:5 (NIV Then the LORD came down in the cloud and stood there with him and proclaimed his name, the LORD. [6] And he passed in front of Moses, proclaiming, "The LORD, the LORD, the compassionate and gracious God, slow to anger, abounding in love and faithfulness,
Ephesians 6:23 (HCSB Peace to the brothers, and love with faith, from God the Father and the Lord Jesus Christ. Colossians 1:3 (HCSB) We always thank God, the Father of our Lord Jesus Christ, when we pray for you,

The Lord Jesus, and our Lord-the Father are two personalities, united in Spirt form (essence), as One God, and Satan denies these two:

2 Corinthians 3:17 (NIV) Now the Lord is the Spirit, and where the Spirit of the Lord is, there is freedom. [18] And we, who with unveiled faces all reflect the Lord's glory, are being transformed into his likeness with ever-increasing glory, which comes from the Lord, who is the Spirit.
Note, only two personalities are denied.
he denies the Father and the Son.

Satan, who is the third personality at the temple (and does not belong there), makes himself out to be god and tempted Christ to worship Him:

> *Again, the devil took him to a very high mountain and showed him all the kingdoms of the world and their splendor. "All this I will give you," he said, "if you will bow down and worship me." Jesus said to him, "Away from me, Satan! For it is written: 'Worship the Lord your God, and serve him only.'" (Matthew 4:8–10, NIV))*

> *Don't let anyone deceive you in any way, for [that day will not come] until the rebellion occurs and the man of lawlessness is revealed, the man doomed to destruction. He will oppose and will exalt himself over everything that is called God or is worshiped, so that he sets himself up in God's temple, proclaiming himself to be God. (2 Thessalonians 2:3–4, NIV)*

Note that sometimes archangels are addressed as "man," as the antichrist (Satan) seems to be addressed as the "man of lawlessness."

> *While I was still in prayer, Gabriel, the man I had seen in the earlier vision, came to me in swift flight about the time of the evening sacrifice. He instructed me and said to me, "Daniel, I have now come to give you insight and understanding. As soon as you began to pray, an answer was given, which I have come to tell you, for you are highly esteemed. Therefore, consider the message and understand the vision. (Daniel 9:21–23, NIV)*

God and the Father make up the temple. And because we have Christ and the Father in us, we too are part of that temple. Satan is not part of the temple.

Gods true temple:

And I saw no temple therein: **for the Lord God Almighty and the Lamb are the temple** *of it. And the city had no need of the sun, neither of the moon, to shine in it: for the* **glory of God did lighten it**, *and* **the Lamb is the light thereof.** *(Revelation 21:22–23, KJV)*

Don't you know that **you yourselves are God's temple and that God's Spirit lives in you**? *If anyone destroys God's temple, God will destroy him; for God's temple is sacred, and you are that temple. (1 Corinthians 3:16–17, NIV)*

Christ and the Father are in us in spiritual form, and that is how we become part of temple.
Romans 8:8–11 JW Bible (proper gender used for Spirit)

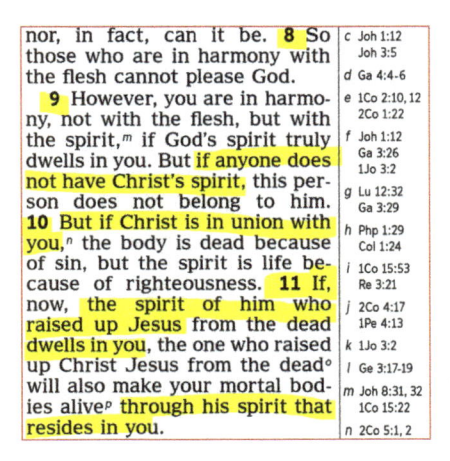

> nor, in fact, can it be. **8 So those who are in harmony with the flesh cannot please God.**
> **9** However, you are in harmony, not with the flesh, but with the spirit,ᵐ if God's spirit truly dwells in you. But **if anyone does not have Christ's spirit**, this person does not belong to him. **10 But if Christ is in union with you**,ⁿ the body is dead because of sin, but the spirit is life because of righteousness. **11** If, now, **the spirit of him who raised up Jesus from the dead dwells in you**, the one who raised up Christ Jesus from the deadᵒ will also make your mortal bodiesᵖ alive **through his spirit that resides in you.**

c Joh 1:12 / Joh 3:5
d Ga 4:4-6
e 1Co 2:10, 12 / 2Co 1:22
f Joh 1:12 / Ga 3:26 / 1Jo 3:2
g Lu 12:32 / Ga 3:29
h Php 1:29 / Col 1:24
i 1Co 15:53 / Re 3:21
j 2Co 4:17 / 1Pe 4:13
k 1Jo 3:2
l Ge 3:17-19
m Joh 8:31, 32 / 1Co 15:22
n 2Co 5:1, 2

What an ingenious deception Satan has come up with. By personifying the Holy Spirit, for Trinitarians, it is the Holy Spirit person who is the Father of Jesus. *Luke 1:35 (NIV): "The angel answered, 'The Holy Spirit will come upon you, and the power of the Most High will overshadow you. So the holy one to be born will be called the Son of God.'"*

And for Trinitarians, it is the Holy Spirit person that raised Jesus from the dead and not the Father! *Romans 8:11 (NIV): "And if the Spirit of him who raised Jesus from the dead is living in you, he who raised Christ from the dead will also give life to your mortal bodies through his Spirit, who lives in you."*

And wasn't Satan, the Antichrist, in a way telling Christ that he was His father by asking for worship? Brilliant liar.

And this Trinity idea started even before Christianity.

The borrowed notion of the Trinity

In the year 2004, my wife and I visited a thousand-year-old temple in Bali and were surprised as they made a presentation about their three-colored trinity of the Buddhist god. When we came back to Calgary, we were shocked to find this same premise creeping into our church. This same lie about a three-colored Trinity Godhead was being preached in our congregation. The people doing it were not evil in my mind, just deceived, and my wife and I decided to leave. This took three more years, but finally, one morning I was informed we needed to advance the three-colored ministry of God, and I could not. That is not what God is.

What God is and what He is doing is hidden in plain sight! The wonderful mystery of what God is and what He has planned for His sons is being desperately squashed and buried by the great deceiver.

Even as ubiquitous as space is, most Christians think of "heaven" as some mysterious place, and not the glorious universe. But what I have been showing from the word of God is that heaven is the glorious star-filled universe—heavens! And shouldn't it be obvious what human families are a type of, yet we don't realize how much human families point to what God is doing—increasing His spiritual family. God is sharing His existence, bringing us into His family (much like the fathers in our human families), yet who knows that? I think it is right there out in the open and obvious. Space is everywhere, and science has helped us see it so much better. And everyone comes from a human father and mother. So why don't we see it?

God will share the glory of Christ and of the universe with adopted sons,—us. And He will be our Father. We will change from physical to spiritual and see God as He truly is, but not yet.

This chapter is to make us aware of how much effort Satan has put into hiding this. And his greatest weapon: the notion of a Trinity.

I say again for effect. Even though ubiquitous space and billions of human families are everywhere before our eyes, we don't see what God is doing with us? He is about family, His family, where we gain the Father's name and our new home that is everywhere (the glorious heavens). We are adopted into that family.

The evil one is a spirit being that wants our worship. He hides our position in that family (like a bird picks up the seed on a path before it can grow) and puts himself in the temple of God and promotes himself to be God (that he is not). I propose that God allows this deception temporarily so we can all be called at the best time for us. The evil one's methods what and how he hides this message, or steals it away, is shown below:

Parable of the sower (rehearsal):

> *He replied, "The knowledge of the secrets of the kingdom of heaven has been given to you, but not to them. Whoever has will be given more, and he will have an abundance. Whoever does not have, even what he has will be taken from him. This is why I speak to them in parables:*
>
> *"Though seeing, they do not see;*
> *though hearing, they do not hear or understand. (Matthew 13:11–13, NIV)*
>
> *"Listen then to what the parable of the sower means: [19] When anyone hears the message about the kingdom and does not understand it, the evil one comes and snatches away what was sown in his heart. This is the seed sown along the path." (Matthew 13:18–19, NIV)*

The agenda for man, as inspired by Satan, to change the word of God is done right out in the open and in plain sight. But just like in the parable of the sower, nobody sees it, not until God opens our eyes. It is my prayer that I can be the dust (soil) that God uses to mix with his spittle (living water) and place upon your eyes (John 9).

The Spirit essence is not a person—it is something that is poured out onto God's people (Israel). In the New Testament, when the apostles were first given their marching orders, Christ told them they would be persecuted and publicly questioned. He told them not to worry about how to answer because the Father's Spirit would speak for them! This was the Father speaking using the power of the Holy Spirit that He is made of.

> *But when they arrest you, do not worry about what to say or how to say it. At that time you will be given what to say, [20] for it will not be you speaking,* ***but the Spirit of your Father speaking through you***. *(Matthew 10:19–20, NIV)*

The Spirit of the Father person, not the Spirit of the Holy Spirit person (there is no such person). Again, the Holy Spirit, is the water like essence (life of God) that flows into us, and gives us life and knowledge.

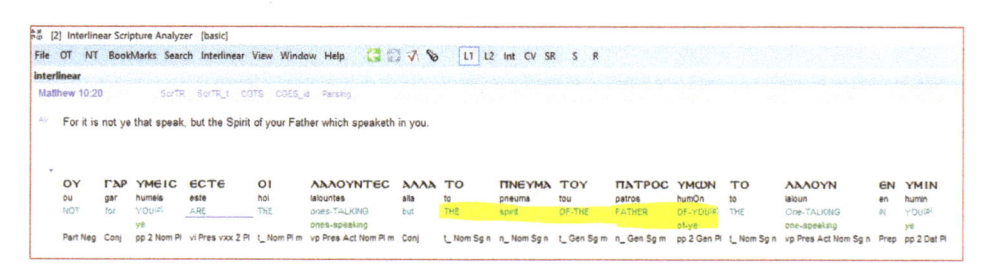

Figure 17. Spirit of the Father (ISA)

In the Old Testament, the "spirit" is shown to be an extension of God, the very power and will of God in action. Spirit is what God is composed of (John 4:24, 2 Corinthians 3:17), and it pours out to accomplish God's will. It is God doing it, but by the extension of the power of His will, His liquid-like Spirit that springs out of him accomplishes it.

> *O LORD, the hope of Israel,*
> *all who forsake you will be put to shame.*
> *Those who turn away from you will be written in the dust*
> *because they have forsaken* **the LORD,**
> **the spring of living water.** *(Jeremiah 17:13, NIV)*
>
> *Jesus answered her, "If you knew* **the gift of God** *and who it is that asks you for a drink, you would have asked him and* **he would have given you living water."** *(John 4:10, NIV)*

Surprisingly, the "spirit" is referred to in the feminine gender in the Old Testament! I believe the Spirit of God has not changed, and I also say that it has not gone from some feminine person to male person across the testaments. *Spirit* is neuter in the New Testament, but I show the scriptures below to point out that Hebrew allows for a female gender pronoun in the Old Testament. Notwithstanding that, the spirit in the Old Testament is never shown to be a person and is very fluid-like, power-like, and will-like. This is truly the same in the New Testament, for God does not change.

Notice as shown in the Interlinear Scripture Analyzer, the green words from (Isaiah 11:2) "and she-rests on him spirit-of Yahweh"

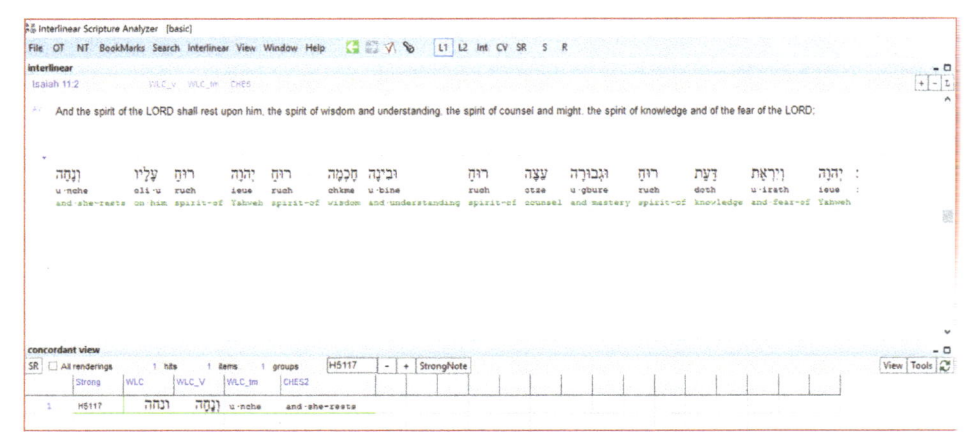

Figure 18. Feminine Gender and the Holy Spirit (ISA)

And from the NIV, which is more readable, no hint of "she." Same thing is true in all translations—no "she."

> *The Branch From Jesse*
> *A shoot will come up from the stump of Jesse;*
> *from his roots a Branch will bear fruit.*
> *The Spirit of the LORD will rest on him—*
> *the Spirit of wisdom and of understanding,*
> *the Spirit of counsel and of power,*
> *the Spirit of knowledge and of the fear of the LORD.*
> *(Isaiah 11:1–2, NIV)*

Below is shown using the Interlinear Analyzer (Ezekiel 2:2), where again, God's Spirit is called she. *"And she is coming in me spirit."* Note that even though the Spirit is feminine, it is He God who is doing the speaking, and it is He God who lifts Ezekiel up—definitely (for me) showing that the Spirit is an extension of God, not a separate person. Otherwise, the so-called third person of the Trinity Godhead is a woman in the Old Testament.

It is true that this is showing a *personification* of the Spirit, or perhaps I should say a *she-onification* in the Old Testament. The Comforter of the New Testament is also a *personification* of the Spirit, or should I say, a *he-onification* in John 14, 15 and 16 (as discussed in chapter 5 in detail).

This feminine attribute of the Spirit in the Old Testament was quite unexpected to me, and I didn't find out until a respected pastor friend of mine (Colin Wallace) pointed it out on the way to services one day in 2016. In keeping with honest analysis, I have included all these facts because I want to convey God's image, not my personal image, in this book. That would be idolatry.

It is also true that the Spirit in the New Testament is always neuter. And because I know God is the same today and forever, it must be some Hebrew language thing, which is not the case in other languages, like Greek, in the New Testament. What I mean is, if the Holy Spirit of God is not feminine in the New Testament, then I don't think that is the meaning God wanted to portray. I don't think God is saying the Holy Spirit is a woman in the biblical intent because she isn't one in the New Testament, and hopefully you all realize by now most importantly she is not a woman in anytime because the Holy Spirit is **not** a person ever.

The book of Ezekiel gives us insight on what Spirit is and who uses it:

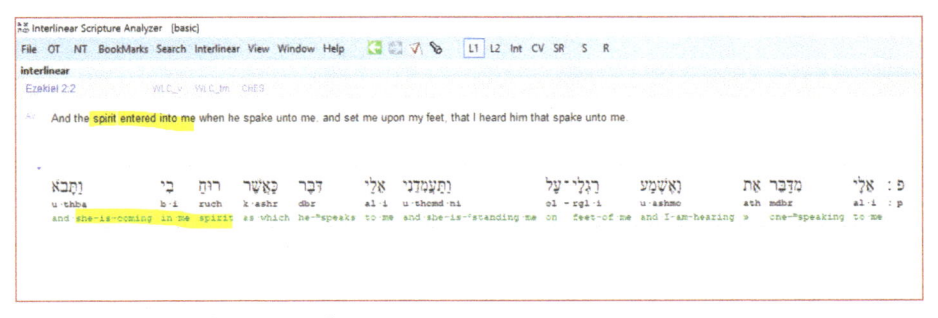

Figure 19. Holy Spirit (She) is coming into me (ISA)

It is interesting to point out that the spirit of animals (creatures) in these Hebrew verses is different from God's Spirit and is not given a feminine attribute. And it is the spirit of the living creature animals that makes the wheels work, "spirit of the animal in the wheels."

Of particular importance, I hope we can see that the spirit is not a different animal. It is the power that comes from the animal (the creature's will-spirit) that makes the wheels turn. Just like with God, it is Him doing things using His Spirit, not some other spirit person.

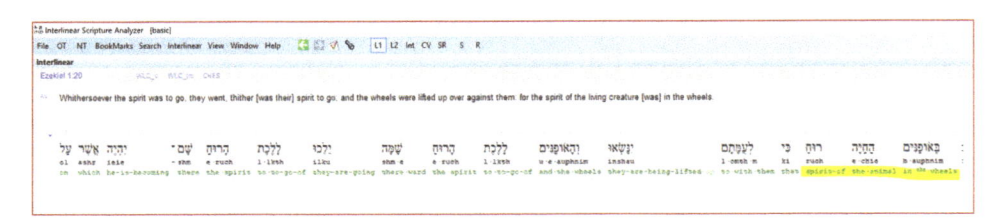

Figure 20. Spirit of the animal, not a different animal, but an extension (ISA)

The spirit, or Spirit, depending on whether it is a lesser living being or God, is the actioning enabler or the will of the being. *It is the spirit, or breath of God, that gives any living thing life. Spirit is life. That is why man, try as he may, cannot create life! But life/spirit can be passed along through the means God gives. Through living seeds, or spiritually, through the laying on of hands.* And it is the being that is really doing it, through the enabling of their spirit/Spirit.

The creatures, or animals, moved about by actioning the wheels spiritually. Similarly, as the wheels move, when our physical hands move, are they not ordered to by the spirit in man? For God breathed a tiny living spirit portion of Himself into us and combined us with a bit of His frozen self (matter dirt), and we became living souls. Complicated world, isn't it? Wow. No wonder man cannot create life, for that would be to create God.

> *But it is the spirit in a man,*
> *the breath of the Almighty, that gives him understanding. (Job 32:8, NIV)*

> *The Spirit of God has made me;*
> *the breath of the Almighty gives me life. (Job 33:4, NIV)*

> *And the LORD God formed man of the dust of the ground, and breathed into his nostrils the breath of life; and man became a living soul. (Genesis 2:7, KJV)*

Just one breath of God's personal Spirit in us brought the dust we are to life as a physical soul (man).

> *The soul that sinneth, it shall die. The son shall not bear the iniquity of the father, neither shall the father bear the iniquity of the son: the righteousness of the righteous shall be upon him, and the wickedness of the wicked shall be upon him. (Ezekiel 18:20, KJV)*

> *Have I any pleasure at all that the wicked should die? saith the Lord GOD: and not that he should return from his ways, and live? (Ezekiel 18:24, KJV)*

> *For the wages of sin is death, but the gift of God is eternal life in Christ Jesus our Lord. (Romans 6:23, HCSB)*

We can play with energy but cannot create it, and we can play with life, but we cannot create it.

God is spirit, God is light, God is love, and God is life (as His word is spirit/life John 6:63).

Getting back to the liquidity of spirit, I believe the Spirit is poured out on God's people, Israel, in both testaments (both the physical Israel of old, and spiritual Israel of the new, over comers and rulers with God.[20]

Please remember, this Old Testament revelation of God and His Spirit **applies to Christians today**, like in Isaiah 9 speaking of a future Jesus, where the prophecy of Israel's salvation comes from: "For unto us a Son is born—wonderful counselor, the mighty God." And so God is the same in both testaments, unchanging.

Remnants of that old covenant Israel are still around, but things have moved on to a new covenant for mankind, and so must the remnant. There is the physical city Jerusalem still, but there will be the spiritual new Jerusalem that the plan of God will continue through uniquely. Promised to Abraham, there is also the remnant of the greatest nation on earth and the remnant of a company of nations, but they are diminishing fast (Genesis 35:11).

We have the same Messiah, for there is only one Messiah—Christ!

> *But **the Messiah has appeared**, **high priest** of the good things that have come. In the greater and more perfect tabernacle not made with hands (that is, not of this creation), He entered the most holy place once for all, not by the blood of goats and calves, but by His own blood, having obtained eternal redemption. For if the blood of goats and bulls and the ashes of a young cow, sprinkling those who are defiled, sanctify for the purification of the flesh, how much more will the blood of the Messiah, who through the eternal Spirit offered Himself without blemish to God, cleanse our consciences from dead works to serve the living God?*
> *Therefore, **He is the mediator of a new covenant**, so that **those who are called might receive the promise** of the eternal inheritance, because a death has taken place for redemption from the transgressions committed under the first covenant. Where a will exists, the death of the one who made it must be established. For a will is valid only when people die, since it is never in force while the one who made it is living. That is why even the first covenant was inaugurated with blood. For when every command had been proclaimed by Moses to all the people according to the law, he took the blood of calves and goats, along with water, scarlet wool, and hyssop, and sprinkled the scroll itself and all the people, saying, **This is the blood of the covenant that God has commanded for you.** In the same way, he sprinkled the tabernacle and all the articles of worship with blood. According to the law almost everything is purified with blood, and without the shedding of blood there is no forgiveness.*

> *Therefore it was necessary for **the copies of the things in the heavens** to be purified with these sacrifices, but the heavenly things themselves to be purified with better sacrifices than these. For the Messiah did not enter a sanctuary made with hands (only a model of the true one) but into heaven itself, so that He might now appear in the presence of God for us. He did not do this to offer Himself many times, as the high priest enters the sanctuary yearly with the blood of another. Otherwise, He would have had to suffer many times since the foundation of the world. **But now He has appeared one time, at the end of the ages, for the removal of sin** by the sacrifice of Himself. And just as it is appointed for people to die once—and after this, judgment—so also the Messiah, having been offered once to bear the sins of many, will appear a second time, not to bear sin, but to bring salvation to those who are waiting for Him.* (Hebrews 9:11–28, HCSB)

The physical remnants of the Old Testament (the "copies of the real thing"), like the twenty-first-century country of Israel, and those that still live in city of Jerusalem must eventually join into the new manifestation of the vine of Israel. The old Jerusalem is only a cherished memory or a favorite textbook from our school days. We have graduated to Christ Jesus, who is now the only door to salvation. The New Jerusalem is the only lasting city. There will be no new Messiah for the physical country of Israel, for there is only one Messiah and He has already accomplished much and will be returning to finish the job. Some are still blindly waiting for Him to come the first time.

> *The Remnant of Israel*
> *I ask then: **Did God reject his people? By no means! I am an Israelite myself, a descendant of Abraham**, from the tribe of Benjamin. God did not reject his people, whom he foreknew. Don't you know what the Scripture says in the passage about Elijah—how he appealed to*

God against Israel: "Lord, they have killed your prophets and torn down your altars; I am the only one left, and they are trying to kill me"? And what was God's answer to him? "I have reserved for myself seven thousand who have not bowed the knee to Baal." So too, at the present time there is a remnant chosen by grace. And if by grace, then it is no longer by works; if it were, grace would no longer be grace.
What then? What Israel sought so earnestly it did not obtain, but the elect did. The others were hardened, as it is written:

**"God gave them a spirit of stupor,
eyes so that they could not see
and ears so that they could not hear,
to this very day."**
And David says:

*"May their table become a snare and a trap,
a stumbling block and a retribution for them.
May their eyes be darkened so they cannot see,
and their backs be bent forever."*

Ingrafted Branches
Again I ask: **Did they stumble so as to fall beyond recovery? Not at all!** *Rather, because of their transgression, salvation has come to the Gentiles to make Israel envious. But if their transgression means riches for the world, and their loss means riches for the Gentiles,* **how much greater riches will their fullness bring!**
I am talking to you Gentiles. Inasmuch as I am the apostle to the Gentiles, I make much of my ministry in the hope that I may somehow arouse my own people to envy and save some of them. For if their rejection is the reconciliation of the world, what will their acceptance be but life from the dead? If the part of the dough offered as firstfruits is holy, then the whole batch is holy; if the root is holy, so are the branches.

> ***If some of the branches have been broken off, and you, though a wild olive shoot, have been grafted in*** among the others and now share in the nourishing sap from the olive root, [18] do not boast over those branches. If you do, consider this: You do not support the root, but the root supports you. You will say then, "Branches were broken off so that I could be grafted in." Granted. But they were broken off because of unbelief, and you stand by faith. Do not be arrogant, but be afraid. For **if God did not spare the natural branches, he will not spare you either**.
> Consider therefore the kindness and sternness of God: sternness to those who fell, but kindness to you, provided that you continue in his kindness. **Otherwise, you also will be cut off. And if they do not persist in unbelief, they will be grafted in, for God is able to graft them in again**. After all, if you were cut out of an olive tree that is wild by nature, and contrary to nature were grafted into a cultivated olive tree, how much more readily will these, the natural branches, be grafted into their own olive tree!
>
> ### All Israel Will Be Saved
> I do not want you to be ignorant of this mystery, brothers, so that you may not be conceited: Israel has experienced a hardening in part until the full number of the Gentiles has come in. And so all Israel will be saved, as it is written:
> > **"The deliverer will come from Zion;**
> > **he will turn godlessness away from Jacob.**
> > **And this is my covenant with them**
> > **when I take away their sins."**
>
> (Romans 11:1–27, NIV)

Please note this is after the Gentiles have been grafted in; it is not speaking of a time when there was only the first covenant with physical-ancient-failed and divorced Israel.

There is only one name under heaven, through which all mankind (Israel) can have their sins removed and be saved—through

Jesus the "I Am." Israel, including Gentiles grafted in, is now living (restored, or grafted, in as Gentiles) in the new or second covenant. The old was divorced and has passed away. Israel, the sons of Abraham having God's eternal promises, has moved on into becoming a new bride of Christ. Eventually Israel- will be spiritual Israel, at-one with God the Father and Christ.

And (for all time), *Galatians 3:29 (NIV)*<u>*: "If you belong to Christ, then you are* ==Abraham's seed==*, and heirs according to the promise.*</u>*"* (Can you only belong to Christ if you are an actual physical descendant of Abraham? No. This shows that all Christians are Abraham's seed. We become part of the people of the promise (Israel—the seed of Abraham) when we became part of Christ!.

> *For I tell you that Christ has become a servant of the Jews on behalf of God's truth,* **to confirm the promises made to the patriarchs** *[9] so that the Gentiles may glorify God for his mercy, as it is written:*
> *"Therefore I will praise you among the Gentiles;*
> *I will sing hymns to your name."*
> *Again, it says,*
> *"Rejoice, O Gentiles, with his people."*
> *[11] And again,*
> *"Praise the Lord, all you Gentiles,*
> *and sing praises to him, all you peoples."*
> *[12] And again,* **Isaiah says,**
> **"The Root of Jesse will spring up,**
> **one who will arise to rule over the nations;**
> **the Gentiles will hope in him."**
> *[13] May the God of hope fill you with all joy and peace as you trust in him,* <u>**so that you may overflow with hope by the power of the Holy Spirit**</u>*. (Romans 15:8–13, NIV)*

In the New Testament, Israel is compared to an olive tree and is shown to include gentiles (wild olive branches), thereby showing that all are Israel, and descendants of the promises which will never fail.

> *After all, if you were cut out of an olive tree that is wild by nature, and contrary to nature were grafted into a cultivated olive tree,* **how much more readily will these, the natural branches, be grafted into their own olive tree!**
> **All Israel Will Be Saved**
> *I do not want you to be ignorant of this mystery, brothers, so that you may not be conceited: Israel has experienced a hardening in part until the full number of the Gentiles has come in.* **And so all Israel will be saved, as it is written:**
> *"The deliverer will come from Zion;*
> *he will turn godlessness away from Jacob.*
> *And this is my covenant with them*
> **when I take away their sins."** *(Romans 11:24–27, NIV)*

Speaking of the **future of us, Israel**:

> *And* **Jesus said** *unto them, Verily I say unto you, that ye which have followed me, in the regeneration* **when the Son of man shall sit in the throne of his glory, ye also shall sit upon twelve thrones, judging the twelve tribes of Israel.** *(Matthew 19:28, KJV)*

The future temple of God has twelve gates that we all will walk through. They have pillars with the names of the twelve apostles and, on the gates, the names of the twelve tribes of Israel!

> *It had a great, high wall with twelve gates, and with twelve angels at the gates.* **On the gates were written the names of the twelve tribes of Israel.** *There were three gates on the east, three on the north, three on the south and three on the west. The wall of the city had twelve foundations,* **and on them were the names of the twelve apostles of the Lamb.** *(Revelation 21:12–14 NIV)*

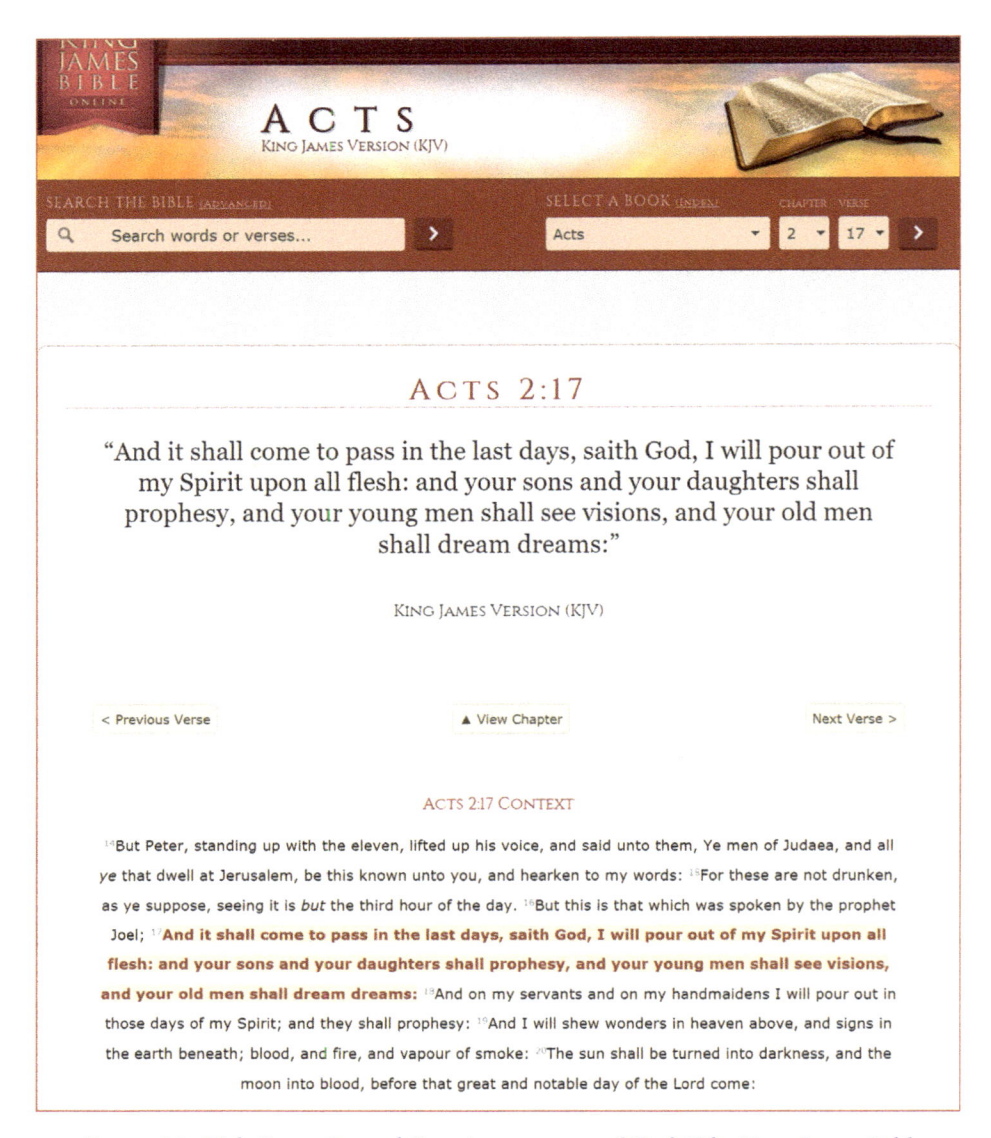

Figure 21. Holy Spirit Poured Out An extension of God (The King James Bible Online)

The Day of the LORD *"And afterward, <u>I will pour out my Spirit on all people</u>.*
Your sons and daughters will prophesy,
your old men will dream dreams,
your young men will see visions.

[29] Even on my servants, both men and women,
I will pour out my Spirit *in those days.*
(Joel 2:28–29, NIV)

When God is pouring out his Spirit (rivers of living water), He is not talking about a third person, He is talking about an extension of Himself (His will etc. John 7:37–39).

Note also, that the spirit spoken of in Joel is the same spirit in the New Testament.

Israel the Chosen
"But now listen, O Jacob, my servant,
Israel, whom I have chosen.
This is what the LORD says—
he who made you, who formed you in the womb,
and who will help you:
Do not be afraid, O Jacob, my servant,
Jeshurun, whom I have chosen.
For I will pour water on the thirsty land,
and streams on the dry ground;
I will pour out my Spirit on your offspring,
and my blessing on your descendants.
They will spring up like grass in a meadow,
like poplar trees by flowing streams.
One will say, 'I belong to the LORD';
another will call himself by the name of Jacob;
still another will write on his hand, 'The LORD's,'
and will take the name Israel. *(Isaiah 44:1–5, NIV)*

So he said to me, "This is the word of the LORD to Zerubbabel: 'Not by might nor by power, but by my Spirit,' says the LORD Almighty. (Zechariah 4:6, NIV)

And yet I have been ==*full of power by the Spirit of Jehovah*==*, And of judgment, and of might, To declare to Jacob his transgression, And to Israel his sin. (Micah 3:8, YLT)*

And in the New Testament:

> May the God of hope fill you with all joy and peace as you trust in him, so that you may overflow **with hope by the power of the Holy Spirit.** (Romans 15:13, NIV)

> My message and my preaching were not with wise and persuasive words, but **with a demonstration of the Spirit's power**, [5] so that your faith might not rest on men's wisdom, but **on God's power**. (1 Corinthians 2:4–5, NIV)

God is the source as a personality; the Spirit is the power and wisdom (like living water) that comes out of him. *"And to those called--both Jews and Greeks--**Christ the power of God, and the wisdom of God"** (1 Corinthians 1:24, YLT).*

God gives us the Spirit that is in him (note: similarly, we also have a spirit within us and that spirit within us is not a different person.

> With my soul[h5315] have I desired[h183] thee in the night[h3915]; **yea, with my spirit[h7307] within[h7130] me will I seek thee early[h7836]:** for when thy judgments[h4941] are in the earth[h776], the inhabitants[h3427] of the world[h8398] will learn[h3925] righteousness[h6664]. (Isaiah 26:9, KJVEC)

> My soul yearns for you in the night;
> in the morning **my spirit longs for you.**
> When your judgments come upon the earth,
> the people of the world learn righteousness.
> (Isaiah 26:9, NIV)

Some examples showing the spirit is not a person but is the will of God (Spirit). *"Woe to apostate sons, The affirmation of Jehovah! To do counsel, and not from Me, And to spread out a covering__, and not of My spirit__, So as to add sin to sin" (Isaiah 30:1, YLT).*

God is Spirit (John 4:24).[21] I believe this is saying God is made up of, or his essence is (water like), Spirit, and not some third person that is called the Holy Spirit and is a third part of the Godhead. For we must worship him in spirit! *"God is spirit, and his worshipers must worship __in spirit__ and in truth" (John 4:24, NIV).*

God is made of Spirit, and we are changing from physical into Spirit essence, which enables us to join a oneness with God the Father and His Son, Jesus (1 Corinthians 15:35–58).

This spirit enters us through the laying on of hands (which is a down payment as we remain physical), and like water, it flows from Christ on God's throne into us as living water. And out from us, as we lay hands on new believers.

You can see this statement made by Christ (or can you?) in the next three pictures. It is shown in both English (King James Authorized Version and with the interlinear Greek from the International Scripture Analyzer).

May God open your eyes to see it. Or do you insist on keeping faith in an idol? That which was inspired by the Antichrist and inserted by deceived clergy—the evil third spirit person at the temple and very throne of God (but not on it) Satan—who appears as an angel of light (2 Thessalonians 2:3, 4).

> *On the last and greatest day of the Feast, Jesus stood and said in a loud voice, __"If anyone is thirsty, let him come to me and drink.__ Whoever believes in me, as the Scripture has said, __streams of living water will flow from within him." By this he meant the Spirit__, whom those who believed in him were later to receive. Up to that time the Spirit had not been given, since Jesus had not yet been glorified. (John 7:37–39, NIV)*

The streams of living water is the Spirit. And from the original Greek:

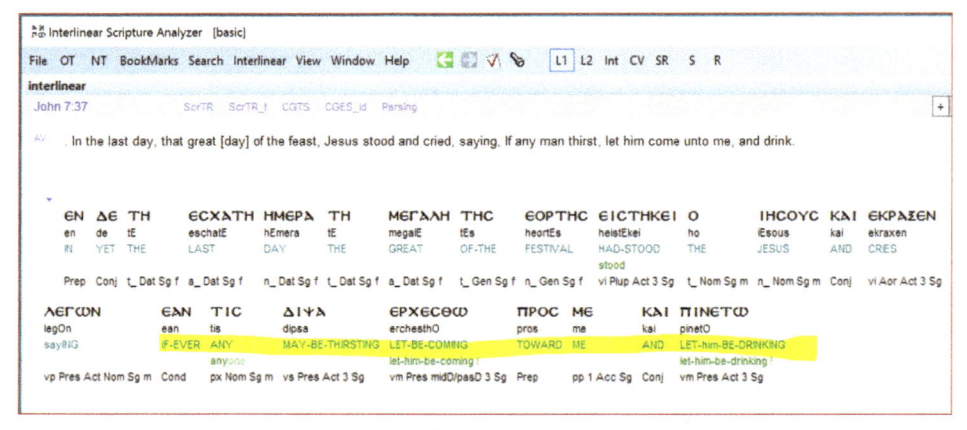

Figure 22. The Streams of Living Water = Holy Spirit (ISA)

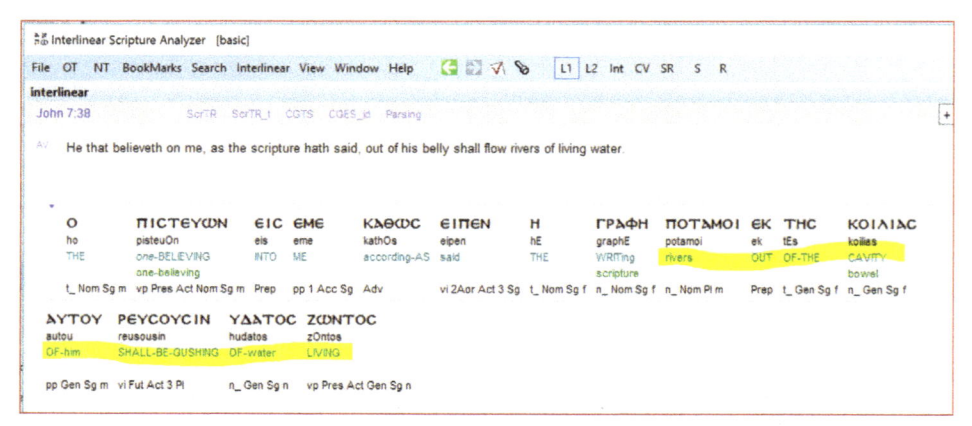

Figure 23. Streams of Living Water Continued (ISA)

SPACE THE TRUE FRONTIER!

Figure 24. Streams of Living Water Continued2 (ISA)

And from the book of Revelation, we see that living-water Spirit originates from the Lamb on the throne with the Father. No third person here, except Satan standing by, acting like he belongs.

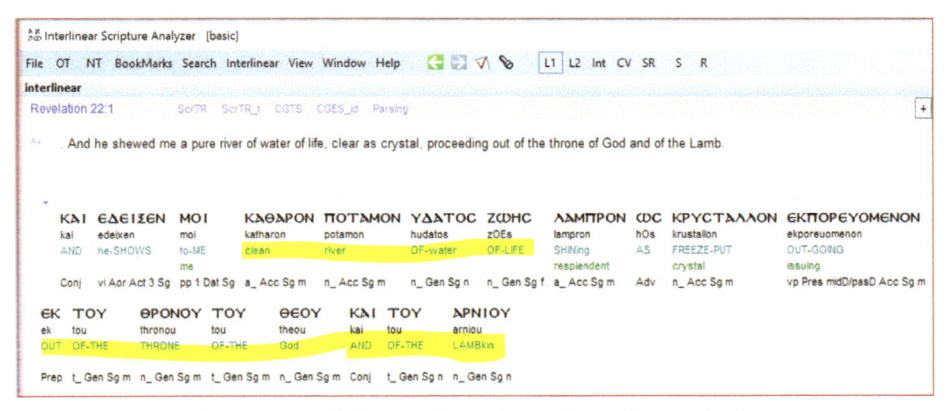

Figure 25. Flowing out of the Lamb on the Fathers Throne (ISA)

If you see this now, I hope you embrace the glory of the truth in thankfulness! And therefore, you may now see why the following statement can be made.

> *The life appeared; we have seen it and testify to it, and* **we proclaim to you the eternal life***, which was* **with the Father and has appeared to us.** *We proclaim to you what we have seen and heard,* **so that you also may**

> *have fellowship with us. And our fellowship **is with the Father and with his Son, Jesus Christ.** We write this **to make our joy complete**. (1 John 1:2–4, NIV)*

The personalities in the Godhead—God the Father and His Son—are a spiritual essence we fellowship with once we are changed to spirit fully. And this **completes** our joy (Father, Son, and bride).

Jesus Prays for All Believers

> "My prayer is not for them alone. **I pray also for those who will believe in me through their message, that all of them may be one, Father, just as you are in me and I am in you. May they also be in us** so that the world may believe that you have sent me. I have given them the glory that you gave me, that they may be one as we are one: I in them and you in me. May they be brought to complete unity to let the world know that you sent me and have loved them even as you have loved me.
>
> "Father, I want those you have given me to be with me where I am, and to see my glory, the glory you have given me because you loved me before the creation of the world.
>
> "Righteous Father, though the world does not know you, I know you, and they know that you have sent me. I have made you known to them, and will continue to make you known in order that the love you have for me may be in them and that I myself may be in them." (John 17:20–25, NIV)

NO THIRD PERSON HERE. Open your eyes to what God shows us—FATHER, SON, AND BRIDE in spiritual essence!

SPACE THE TRUE FRONTIER!

Figure 26. The Spirit of Christ in us (ISA)

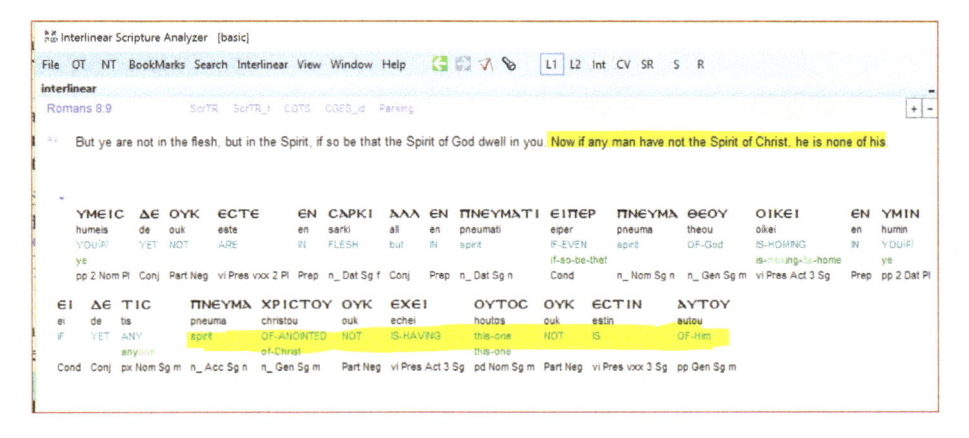

Figure 27. The Spirit of Christ and the Father in us (ISA)

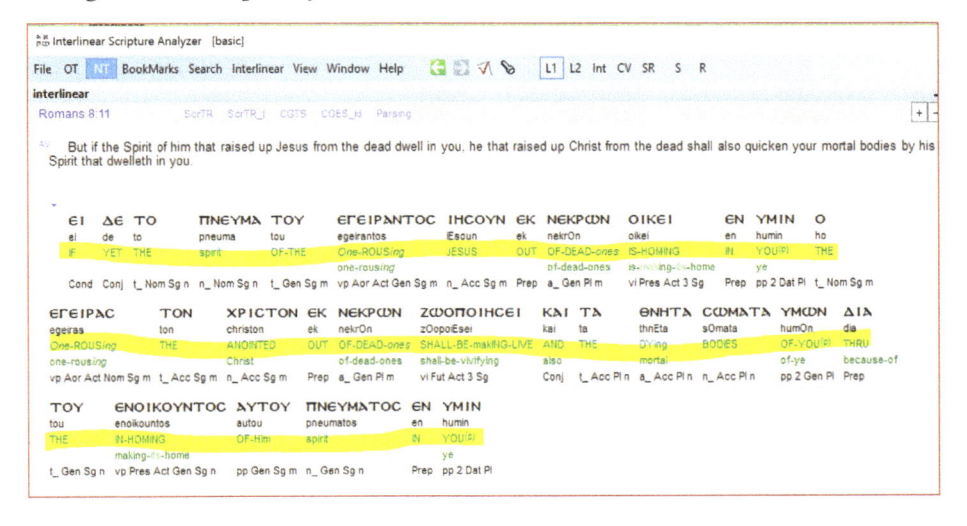

There are two realities of existence: the first is flesh or physical, the better and final is spiritual. It is a fellowship in spiritual unity with God the Father, and Christ dwelling in us.

They are our Comforter! It is Christ and the Father that dwell in us in spirit form—not some third personality. That's why Christ waited until He was resurrected into His former spiritual glory (John 7:39) before he could enter us and come with the Father (combined **Comforter**).

Christ, as the flesh that he was on earth, had to wait until He was restored to His previous glorious spirit state before He could come into us. And when He is in us, the Father is with Him and, therefore, also in us (John 17)—one spirit family of God. Father, Son, and bride! "Oh Be One Can Sow Be," as opposed to "Oh Be One Can No Be."

This scripture below shows that it is Christ and the Father that are the Comforter-Spirit within us.

Romans 8:8–11 from the JW Bible:

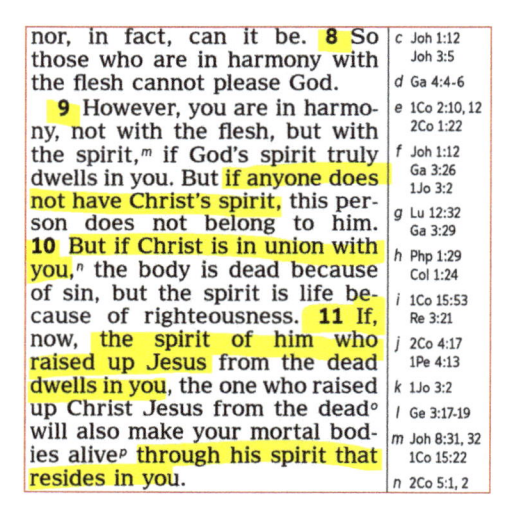

| nor, in fact, can it be. **8** So those who are in harmony with the flesh cannot please God. **9** However, you are in harmony, not with the flesh, but with the spirit,ᵐ if God's spirit truly dwells in you. But **if anyone does not have Christ's spirit,** this person does not belong to him. **10 But if Christ is in union with you,**ⁿ the body is dead because of sin, but the spirit is life because of righteousness. **11** If, now, **the spirit of him who raised up Jesus from the dead dwells in you**, the one who raised up Christ Jesus from the deadᵒ will also make your mortal bodiesᵖ alive **through his spirit that resides in you.** | c Joh 1:12 Joh 3:5 <br> d Ga 4:4-6 <br> e 1Co 2:10, 12 2Co 1:22 <br> f Joh 1:12 Ga 3:26 1Jo 3:2 <br> g Lu 12:32 Ga 3:29 <br> h Php 1:29 Col 1:24 <br> i 1Co 15:53 Re 3:21 <br> j 2Co 4:17 1Pe 4:13 <br> k 1Jo 3:2 <br> l Ge 3:17-19 <br> m Joh 8:31, 32 1Co 15:22 <br> n 2Co 5:1, 2 |

John 14: 17–20 JW Bible (correctly not personifying the Spirit)

Jesus Christ, returned to His former spirtiual glory, can now keep his promise and come to us with the Father (indwelling as Comforter).

I AM IN YOU, WITH MY FATHER! THEY COMFORT US, AS THEY SPIRITUALLY DWELL IN US! HE DIDN'T SEND SOME VICAR (THIRD PERSON). CHRIST SAID, "I AM COMING TO YOU." IT IS CHRIST WHO COMES TO US, AND WITH HIM IS THE FATHER (TO COMPLETE OUR UNITY). IT IS THEIR SPIRITUAL PRESENCE THAT COMFORTS US.

SPACE THE TRUE FRONTIER!

John 14:17-20 JW Bible

> to be with you forever,^c **17 the spirit of the truth,**^d **which the** world cannot receive, because it neither sees it nor knows it.^e You know it, **because it remains with you and is in you. 18 I will not leave you bereaved.* I am coming to you.**^f **19** In a little while the world will see me no more, but you will see me,^g because I live and you will live. **20** In that day you will know that **I am in union with my Father** and you are in union with me and **I am in union with you.**^h **21** Whoever

And the Egyptians are men, and not God, And their horses are flesh, and not spirit, And Jehovah stretcheth out His hand, And stumbled hath the helper, And fallen hath the helped one, And together all of them are consumed. (Isaiah 31:3, YLT)

Showing men are flesh and God is spirit, what we are made of and what God is made of. Eventually, we will all be one in spirit.

Again, the scripture showing what is in us as Comforter is **Christ and the Father**. And from the New World translated SpirIT, showing man also has a spirit in him, that is not another person. So too with God.

God is a Spirit—that is not another third person of the Godhead.

1 Corinthians 2:11–14 (JW Bible New World Translation)

> **11** For who among men knows the things of a man except the man's spirit within him? So, too, no one has come to know the things of God except the spirit of God. **12** Now we received, not the spirit of the world, but the spirit that is from God,ᵉ so that we might know the things that have been kindly given us by God. **13** These things we also speak, not with words taught by human wisdom,ᶠ but with those taught by the spirit,ᵍ as we explain* spiritual matters with spiritual words.

In the Old Testament, there is a good example of what I am saying. **The spirit is not the person!** Elisha is reprimanding Gehazi for taking money from Naaman (who had just been healed of leprosy after following Elisha's directions to wash in the Jordan seven times). Naaman had offered money to Elisha, but he refused it. Gehazi, on the other hand, secretly chased after Naaman and asked a reward. Later, Gehazi denied his journey and lied to Elisha, saying he had gone nowhere, but the spirit or heart or will of Elisha had been present with him. He says, "Don't you know my heart or spirit [inner man, will] was there with you?'.

Notice that from the *NAS Exhaustive Concordance Hebrew Dictionary*, we see the Hebrew word **leb #3820** (based on Hebrew word *lebab*, #3824) is used to reveal that Elisha was present with Gehazi in a different than physical way. That his heart, mind, will, or spirit was with him. In fact, that is how the **NIV** translates this word (as spirit).

*And he said[h559] unto him, Went[h1980] not mine **heart[h3820]** with thee, when the man[h376] turned[h2015] again from his chariot[h4818] to meet[h7125] thee? Is it a time[h6256] to receive[h3947] money[h3701], and to receive[h3947] garments[h899], and oliveyards[h2132], and vineyards[h3754], and sheep[h6629], and oxen[h1241], and menservants[h5650], and maidservants[h8198]? (2 Kings 5:26, KJV Exhaustive Concordance Version)*

*"By all means, take two talents," said Naaman. He urged Gehazi to accept them, and then tied up the two talents of silver in two bags, with two sets of clothing. He gave them to two of his servants, and they carried them ahead of Gehazi. When Gehazi came to the hill, he took the things from the servants and put them away in the house. He sent the men away and they left. Then he went in and stood before his master Elisha. "Where have you been, Gehazi?" Elisha asked. "Your servant didn't go anywhere," Gehazi answered. But Elisha said to him, **"Was not my spirit with you** when the man got down from his chariot to meet you? Is this the time to take money, or to accept clothes, olive groves, vineyards, flocks, herds, or menservants and maidservants? Naaman's leprosy will cling to you and to your descendants forever." Then Gehazi went from Elisha's presence and he was leprous, as white as snow. (2 Kings 5:23–27, **NIV**)*

> **NAS Exhaustive Concordance Hebrew-Aramaic Dictionary** 3820. leb
>
> 3820. לֵב **leb** (524b); from the same as 3824; *inner man, mind, will, heart*:—accord(1), attention(4), attention*(1), bravest*(1), brokenhearted*(3), care*(2), celebrating*(1), chests*(1), completely*(1), concern*(1), concerned*(1) conscience(1), consider*(2), considered*(2), courage(1), decided*(1), determine*(1), discouraged*(1), discouraging*(1), doing*(1), double heart(1), encouragingly*(1), heart(396), heart's(2), hearts(40), Himself(1), himself(6), imagination(1), inspiration(2), intelligence(1), kindly(5), life(1), merry-hearted*(1), middle(2), midst(1), mind(36), minds(3), myself(6), obstinate*(2), planned*(1), presume*(1), pride*(1), recalls*(1), reflected*(1), regard*(1), self-exaltation*(1), sense(10), senseless*(1), seriously(1), skill*(1), skilled*(1), skillful man*(1), skillful men*(1), skillful persons*(1), skillful*(3), spirits(1), stouthearted*(1), stubborn-minded*(1), tenderly(2), thought(3), understanding(7), undivided*(1), well(2), willingly*(1), wisdom(2), yourself(1), yourselves(1).
>
> 3821. לֵב **leb** (1098d); (Ara.) corr. to 3820; *heart*:—myself(1).
>
> 3822. לְבָאוֹת **Lebaoth** (522d); from the same as 3833b; *a city in S. Judah*:—Lebaoth(1).
>
> 3823a. לָבַב **labab** (525d); denom. vb. from 3824; *to get a mind* or *to encourage*:—become intelligent(1), heart beat faster(2).
>
> 3823b. לָבַב **labab** (525d); denom. vb. from 3834; *to make cakes*:—made cakes(1), make(1).
>
> 3824. לֵבָב **lebab** (523b); from an unused word; *inner man, mind, will, heart*:—anger(1), breasts(1), conscientious*(1), consider*(5), courage(1), desire(1), encouragingly*(1), fainthearted*(3), heart(185), heart and the hearts(1), heart's(1), hearts(27), hearts like his heart(1), intelligence(1), intended(2), mind(8), purpose(1), thought(1), timid*(1), understanding(2), wholehearted*(1), wholeheartedly*(1), yourself(1).

Figure 28. Elisha's spirit, was there with Gehazi, (NAS Exhaustive Concordance Hebrew-Aramaic Dictionary)

The point is that when Elisha's spirit, or heart, was with Gehazi, it was not someone else! The same thing is true when God's Spirit is in you, it is the Father and the Son Jesus, who are spiritually there with you. Not some other person.

The Laying on of Hands: The pathway of the Fluid Spirit

The Spirit of God that moved upon the waters of the earth in the Genesis creation (probably the re-creation) is the same spirit that flows out of the Lamb on the throne of God in the New Testament. And God gives that spirit (His will extension, river of water, and everlasting life) through the laying on of hands, which is one of the foundational doctrines largely ignored nowadays.

> *For this reason I remind you to fan into flame the gift of God, which is in you through the laying on of my hands. [7] For God did not give us a spirit of timidity,* ***but a spirit of power, of love and of self-discipline****. (2 Timothy 1:6–7, NIV)*

That is not a third person; it is the extension of, it is the power of, it is the will of our God.

The spirit (oneness) is the Father and the Son right now. And He has started the process of our birth into his likeness, with the **down payment** of the spirit life of our future, through the laying on of hands. *"Now it is God who makes both us and you stand firm in Christ. He anointed us, [22] set his seal of ownership on us, and put his Spirit in our hearts as* a deposit*, guaranteeing what is to come" (2 Corinthians 1:21–22, NIV)*. We are not fully born into God's family until Christ's return at the last trump, but we have a down payment, or deposit, and are sealed by the spirit of God, guaranteeing our future change.

> *But* you have an anointing from the Holy One*, and all of you know the truth. I do not write to you because you do not know the truth, but because you do know it*

> *and because no lie comes from the truth. Who is the liar? It is the man who denies that Jesus is the Christ. Such a man is the antichrist—**he denies the Father and the Son. No one who denies the Son has the Father; whoever acknowledges the Son has the Father also.** See that what you have heard from the beginning remains in you. If it does, you also will **remain in the Son and in the Father**. And this is what he promised us—even eternal life. (1 John 2:20–25, NIV)*

Please note, we remain in two personalities mentioned, the Father and the Son, not three persons (John 17:20–23). Also, we see Satan denying only two members of the Godhead. Why not three?

The scripture below is about a different Zechariah, which I read about in my daily readings for June 14, 2017. It just fit so well I wanted to include it now. After going through all this political turmoil of our times, I believe we should open our eyes and realize who gave us the greatness we had. We act like it came from our leaders, and that if we just elect the right one, things will get back to the great prosperity and freedoms we once had. We are like Nebuchadnezzar if we believe a man can do this! **The problem is not our leadership—it is us.**

> ***And the Spirit of God came upon Zechariah*** *the son of Jehoiada the priest, which stood above the people, and said unto them, Thus saith God, Why transgress ye the commandments of the LORD, that ye cannot prosper?* ***because ye have forsaken the LORD, he hath also forsaken you.*** *(2 Chronicles 24:20, KJV)*

When I mentioned to a young person, what if the potential leader is lying, she replied, "Who cares as long as she does a good job lying for me?"

Politicians need to start with a lie-detector test, but I'm not sure we could trust the tester.

Not One Is Upright
"Go up and down the streets of Jerusalem,
look around and consider,
search through her squares.
If you can find but one person
who deals honestly and seeks the truth,
I will forgive this city.
Although they say, 'As surely as the LORD lives,'
still they are swearing falsely."
O LORD, do not your eyes look for truth?
You struck them, but they felt no pain;
you crushed them, but they refused correction.
They made their faces harder than stone
and refused to repent.
I thought, "These are only the poor;
they are foolish,
for they do not know the way of the LORD,
the requirements of their God. *(Jeremiah 5:1–4, NIV)*

Today I also happened to read the first two chapters of Ephesians, and it totally came alive for me. I hadn't realized how much of what I am trying to share is found in these chapters. I have decided to insert these chapters for that reason. They say it so well.

Please notice that the "will of God" is talked about in verse 1 (which I am relating to the **thing, Gods spirit is**), and then also notice who's there. Who is this from? It is from Jesus and the Father, delivered gracefully by the will of God and given to us as a down payment of our future change into the full likeness of God (Ephesians 1:13). And please especially notice Ephesians 2:21–22, where you see that we are eventually part of a spiritual unity at one with God—God the Father person, Jesus Christ the Lamb person, and us—our inheritance into the likeness of God. No three-person Trinity. **If the Holy Spirit was a separate person, it would be impossible to leave "it" (and for Trinitarians "Him") out of these two chapters!** Enjoy!

Paul, an apostle of Jesus Christ through the will of God, to the saints who are in Ephesus, and to the faithful in Christ Jesus: Grace to you, and peace from God our Father, and the Lord Jesus Christ!
Blessed is the God and Father of our Lord Jesus Christ, who did bless us in every spiritual blessing in the heavenly places in Christ, according as He did choose us in him before the foundation of the world, for our being holy and unblemished before Him, in love, having foreordained us to the adoption of sons through Jesus Christ to Himself, according to the good pleasure of His will, to the praise of the glory of His grace, in which He did make us accepted in the beloved, in whom we have the redemption through his blood, the remission of the trespasses, according to the riches of His grace, in which He did abound toward us in all wisdom and prudence, having **made known to us the secret of His will**, *according to His good pleasure, that He purposed in Himself, in regard to the dispensation of the fulness of the times, to bring into one the whole in the Christ, both the things in the heavens, and the things upon the earth--in him;* **in whom also we did obtain an inheritance, being foreordained according to the purpose of Him who the all things is working according to the counsel of His will,** *for our being to the praise of His glory, even those who did first hope in the Christ, in whom ye also, having heard the word of the truth--the good news of your salvation--in whom also having believed*, **ye were sealed with the Holy Spirit of the promise,** *which is an earnest of our inheritance, to the redemption of the acquired possession, to the praise of His glory.*
Because of this I also, having heard of your faith in the Lord Jesus, and the love to all the saints, do not cease giving thanks for you, making mention of you in my prayers, **that the God of our Lord Jesus Christ, the Father of the glory, may give to you a spirit of wisdom and**

revelation in the recognition of him, the eyes of your understanding being enlightened, for your knowing what is the hope of His calling, and what the riches of the glory of His inheritance in the saints, and what the exceeding greatness of His power to us who are believing, <u>**according to the working of the power of His might,**</u> **which He wrought in the Christ,** *having raised him out of the dead, and did set him at His right hand in the heavenly places, far above all principality, and authority, and might, and lordship, and every name named, not only in this age, but also in the coming one; and all things He did put under his feet, and did give him--head over all things to the assembly, which is his body, the fulness of Him who is filling the all in all. (Ephesians 1:1–23, YLT)*

Also you--being dead in the trespasses and the sins, in which once ye did walk according to the age of this world, according to the ruler of the authority of the air, of the spirit that is now working in the sons of disobedience, among whom also we all did walk once in the desires of our flesh, doing the wishes of the flesh and of the thoughts, and were by nature children of wrath--as also the others, and God, being rich in kindness, because of His great love with which He loved us, even being dead in the trespasses, did make us to live together with the Christ, **(by grace ye are having been saved,) and did raise us up together, and did seat us together in the heavenly places in Christ Jesus**, *that He might show, in the ages that are coming, the exceeding riches of His grace in kindness toward us in Christ Jesus, for by grace ye are having been saved, through faith, and this not of you--of God the gift, not of works, that no one may boast; for of Him we are workmanship, created in Christ Jesus to good works, which God did before prepare, that in them we may walk.*

> *Wherefore, remember, that ye were once the nations in the flesh, who are called Uncircumcision by that called Circumcision in the flesh made by hands, that ye were at that time apart from Christ, having been alienated from the* **commonwealth of Israel, and strangers to the covenants of the promise**, *having no hope, and without God, in the world; and now, in Christ Jesus, ye being once afar off became nigh in the blood of the Christ, for he is our peace, who did make both one, and the middle wall of the enclosure did break down, the enmity in his flesh, the law of the commands in ordinances having done away, that the two he might create in himself into one new man, making peace, and might reconcile both in one body to God through the cross, having slain the enmity in it, and having come, he did proclaim good news--peace to you--the far-off and the nigh, because through him we have the access--we both--in one Spirit unto the Father. Then, therefore, ye are no more strangers and foreigners, but fellow-citizens of the saints, and of the household of God, being built upon the foundation of the apostles and prophets, Jesus Christ himself being chief corner-stone,* **in whom all the building fitly framed together doth increase to an holy sanctuary in the Lord, in whom also ye are builded together, for a habitation of God in the Spirit.** *(Ephesians 2:1–22, YLT)*

Remember Gods spirit "fluttered" on the ruined earth and repaired it, bringing it out of its former decay. The "fluttering spirit" power of God did this, not some third individual person. **Jesus Christ created everything** through his spiritual power, not a third person. Genesis 1:1–2 (KJV): *"In the beginning God created the heaven and the earth. [2] And the earth was without form, and void; and darkness was upon the face of the deep. And* **the Spirit of God moved upon the face of the waters**.*"*

If you are a Trinitarian, you have a conflict because the Creator would be the third-person Holy Spirit. This contradicts that the Christ person is the Creator.

> *The Supremacy of Christ*
> *He is the image of the invisible God, the firstborn over all creation.* **For by him all things were created: things in heaven and on earth, visible and invisible**, *whether thrones or powers or rulers or authorities; all things were created by him and for him. (Colossians 1:15–16, NIV)*

It is interesting to note that the wave action, or fluttering, is similar to what science is discovering about matter at the quantum level. *Genesis 1:1 (YLT): "In the beginning of God's preparing the heavens and the earth—the earth hath existed waste and void, and darkness is on the face of the deep, and* **the Spirit of God fluttering on the face of the waters.**"

> Jesus returned to Galilee in the power of the Spirit, and news about him spread through the whole countryside. He taught in their synagogues, and everyone praised him.
> He went to Nazareth, where he had been brought up, and on the Sabbath day he went into the synagogue, as was his custom. And he stood up to read. The scroll of the prophet Isaiah was handed to him. Unrolling it, he found the place where it is written:
> "The Spirit of the Lord is on me,
> because he has anointed me
> to preach good news to the poor.
> He has sent me to proclaim freedom for the prisoners
> and recovery of sight for the blind,
> to release the oppressed,
> to proclaim the year of the Lord's favor."
> (Luke 4:14–19, NIV)

> *Therefore I glory in Christ Jesus in my service to God. I will not venture to speak of anything except what Christ has accomplished through me in leading the Gentiles to obey God by what I have said and done—by the power of signs and miracles, **through the power of the Spirit**. So from Jerusalem all the way around to Illyricum, I have fully proclaimed the gospel of Christ. (Romans 15:17–19, NIV)*

> ***But as for me, I am filled with power, with the Spirit of the LORD,***
> *and with justice and might,*
> *to declare to Jacob his transgression,*
> *to Israel his sin. (Micah 3:8, NIV)*

The Father is the personality that anointed Jesus with a purpose and a job. And the Father did it through the means of His Spirit. The Father was in Jesus! Christ was not alone; He and the Father were one spiritually. And it is Christ who is in us, because His Spirit is in us (not a different spirit person, but Christ). And with Christ in us, because the Father is in Him, so is the Father in us. (Father, Son, bride in John17).

> *I have been crucified with Christ and I no longer live, **but Christ lives in me**. The life I live in the body, I live by faith in the Son of God, who loved me and gave himself for me. (Galatians 2:20, NIV)*

> *You, however, are controlled not by the sinful nature but by the Spirit, **if the Spirit of God lives in you. And if anyone does not have the Spirit of Christ**, he does not belong to Christ. (Romans 8:9, NIV)*

Again, it is Christ's spirit that is in us that makes us Christians, not some third spirit person.

Life by the Spirit

Another point is the example from Galatians, admonishing us to live spiritually and not sinfully. **The point I want to make here is that this scripture is talking about "ways of life." It is not talking about people in us. The way of the Spirit is Christ and his attitudes living in us.** These natures come from two personalities, either the source of good (Christ) or the source of evil (man under Satan's way). This "way of the Spirit" is not a third personality! It is Christ that lives in us! (Galatians 2:20)

This scripture is talking about ways of life, and it is the person of Christ that is in us (in spiritual form) that enables us to live spiritually (according to the spirit).

> *And I say: In the Spirit walk ye, and the desire of the flesh ye may not complete;* ***for the flesh doth desire contrary to the Spirit, and the Spirit contrary to the flesh, and these are opposed one to another,*** *that the things that ye may will--these ye may not do; and if by the Spirit ye are led, ye are not under law. And manifest also are the works of the flesh, which are: Adultery, whoredom, uncleanness, lasciviousness, idolatry, witchcraft, hatred, strifes, emulations, wraths, rivalries, dissensions, sects, envyings, murders, drunkennesses, revellings, and such like, of which I tell you before, as I also said before, that those doing such things the reign of God shall not inherit. And the fruit of the Spirit is: Love, joy, peace, long-suffering, kindness, goodness, faith, meekness, temperance: against such there is no law; and those who are Christ's, the flesh did crucify with the affections, and the desires;* ***if we may live in the Spirit, in the Spirit also we may walk****. (Galatians 5:16–25, YLT)*

We walk with Christ's Spirit! Christ in spiritual essence (not some third spiritual person).

Here are other examples, showing that the spirit of God is not a personality but that it is like a flowing life-giving liquid. And they show that God the Father and Christ (the former Logos or Word) are the personalities of the Godhead.

Don't you agree that if there were three, they would all be talked about here? I mean, we are talking about the beginning of something big, and all the stakeholders should be there, right? And that it would be more respectful to say something like, "In the beginning was the Father person, the Holy Spirit person, and the Logos—Word person that became Christ."

What is revealed here about the Godhead is that there are the persons of the Father, and the Logos (future Christ) and they are spiritual. The third person is not omitted; there simply is no third person.

> *The Word Became Flesh*
> *In the beginning was the Word, and the Word was with God, and the Word was God. He was with God in the beginning.*
> *Through him all things were made; without him nothing was made that has been made. In him was life, and that life was the light of men. The light shines in the darkness, but the darkness has not understood it. (John 1:1–5, NIV)*

If God is made of spirit, then what is that like: Christ shows that the spirit of God is like a stream of living water! Christ talks about himself being with the Father, and they are together a oneness made up of spirit essence that they pour out of themselves into us.

> *God is spirit, and his worshipers must worship in spirit and in truth. (John 4:24, NIV)*

> *On the last and greatest day of the Feast, Jesus stood and said in a loud voice, "If anyone is thirsty, let him come to me and drink. Whoever believes in me, as the*

SPACE THE TRUE FRONTIER!

> *Scripture has said, **streams of living water will flow from within him**." **By this he meant the Spirit**, whom those who believed in him were later to receive. Up to that time the Spirit had not been given, since Jesus had not yet been glorified. (John 7:37–39, NIV)*
>
> *Jesus answered her, "If you knew the gift of God and who it is that asks you for a drink, you would have asked him and he would have given you living water." (John 4:10, NIV)*
>
> *O LORD, the hope of Israel,*
> *all who forsake you will be put to shame.*
> *Those who turn away from you will be written in the dust because they have forsaken **the LORD,***
> ***the spring of living water.** (Jeremiah 17:13, NIV)*
>
> *Then the angel showed me the river of the water of life, as clear as crystal, **flowing from the throne of God and of the Lamb** down the middle of the great street of the city. On each side of the river stood the tree of life, bearing twelve crops of fruit, yielding its fruit every month. And the leaves of the tree are for the healing of the nations. (Revelation 22:1–22, NIV)*

Note that the spirit is flowing out of the Lamb, who is undeniably Christ. It is the river of life, and note who is on the throne! The Father and the Lamb. So where is the third person?

> **The Day of the LORD**
> *"And afterward, **I will pour out my Spirit on all people**.*
> *Your sons and daughters will prophesy,*
> *your old men will dream dreams,*
> *your young men will see visions.*
> *Even on my servants, both men and women,*
> ***I will pour out my Spirit** in those days.*
> *(Joel 2:28–29, NIV)*

The spirit of God is his liquid-like essence. It is what He is made of, and it "flutters" wave-like in its action.

Remember, as we are temporarily made of flesh, eventually we will be born again into the very spirit likeness of God. We will then be made of spirit, just like the Father and Christ.

> *So will it be with the resurrection of the dead. The body that is sown is perishable, it is raised imperishable; it is sown in dishonor, it is raised in glory; it is sown in weakness, it is raised in power; **it is sown a natural body, it is raised a spiritual body. If there is a natural body, there is also a spiritual body.** So it is written: "The first man Adam became a living being"; the last Adam, a life-giving spirit. The spiritual did not come first, but the natural, and after that the spiritual. The first man was of the dust of the earth, the second man from heaven. (1 Corinthians 15:42–47, NIV)*

We do not come into the fullness of the likeness of God, until we are born a second time as spirit beings. The next scripture shows what happens to those who die as Christian's (are asleep) at Christ's return to earth as King (for the trumpet shall sound, and the dead in Christ shall be raised.

> *I declare to you, brothers, that flesh and blood cannot inherit the kingdom of God, nor does the perishable inherit the imperishable. Listen, **I tell you a mystery: We will not all sleep, but we will all be changed**—in a flash, in the twinkling of an eye, at the last trumpet. For the trumpet will sound, the dead will be raised imperishable, and we will be changed. For the perishable must clothe itself with the imperishable, and the mortal with immortality. When the perishable has been clothed with the imperishable, and the mortal*

> *with immortality, then the saying that is written will come true: "Death has been swallowed up in victory." (1 Corinthians 15:50–54, NIV)*

> ***"At that time the sign of the Son of Man will appear in the sky**, and all the nations of the earth will mourn. They will see the Son of Man coming on the clouds of the sky, with power and great glory. And **he will send his angels with a loud trumpet call**, and they **will gather his elect** from the four winds, from one end of the heavens to the other. (Matthew 24:30–31, NIV)*

And then we shall be like God, just as he said, spiritual ever-living adopted sons of God!

Those who are fortunate enough partake of the "tree of life," which is fed by the living water that flows from the Lamb on Gods throne, and written in the book of life. On that throne are God the Father, and the Lamb (Christ), and spirit life flows out to us.

> *The River of Life*
> *Then the angel showed me the river of the water of life, as clear as crystal, flowing from the throne of God and of the Lamb down the middle of the great street of the city. On each side of the river stood the tree of life, bearing twelve crops of fruit, yielding its fruit every month. And the leaves of the tree are for the healing of the nations. No longer will there be any curse. The throne of God and of the Lamb will be in the city, and his servants will serve him. They will see his face, and his name will be on their foreheads. There will be no more night. They will not need the light of a lamp or the light of the sun, for the Lord God will give them light. And they will reign for ever and ever. (Revelation 22:1–5, NIV)*

He who has an ear, let him hear what the Spirit says to the churches. **To him who overcomes, I will give the right to eat from the tree of life,** *which is in the paradise of God. (Revelation 2:7, NIV)*

He who overcomes will, like them, be dressed in white. ***I will never blot out his name from the book of life, but will acknowledge his name before my Father and his angels****. He who has an ear, let him hear what the Spirit says to the churches. (Revelation 3:5–6, NIV)*

And who is that Spirt that is speaking? **And why does Christ only show this to the Father and his angels if there is some other third person in the Godhead?**

> ***Now the Lord is the Spirit****, and where the Spirit of the Lord is, there is freedom. And we, who with unveiled faces all reflect the Lord's glory, are being transformed into his likeness with ever-increasing glory,* ***which comes from the Lord, who is the Spirit****. (2 Corinthians 3:17–18, NIV)*

I want to get back to revealing the agenda to deceive mankind. One of the most obvious areas where mankind has agreed to the guiles of Satan is in deliberately mistranslating one tiny word—it is an "IT"! It is a discredit to those who have mistranslated the scriptures, changing the gender of Spirit from an "it" to a "he." I question your honesty if you deny this after I show the obvious below.

There are hundreds of places in the New Testament where the neuter gender-spirit is incorrectly given a male gender to fit the image of god that they want (inspired by the great imposter god that deceives the whole world).

It is noteworthy and a credit to the translators of the Jehovah Witnesses version of the Bible (New World Translation) that kept true to the intent of God and kept the gender neuter.

SPACE THE TRUE FRONTIER!

Jesus Promises the Holy Spirit
"If you love me, you will obey what I command. And I will ask the Father, and he will give you another Counselor to be with you forever—the Spirit of truth. ==The world cannot accept him, because it neither sees him nor knows him==*. But you know him, for he lives with you and will be in you. [18] I will not leave you as orphans; I will come to you. Before long, the world will not see me anymore, but you will see me. Because I live, you also will live. On that day you will realize that I am in my Father, and you are in me, and I am in you. Whoever has my commands and obeys them, he is the one who loves me. He who loves me will be loved by my Father, and I too will love him and show myself to him."*
(John 14:15–21, NIV)

14:17	TO to G3588 t_ Acc Sg n THE	ΠΝΕΥΜΑ pneuma G4151 n_ Acc Sg n spirit	ΤΗΣ tEs G3588 t_ Gen Sg f OF-THE	ΑΛΗΘΕΙΑΣ alEtheias G225 n_ Gen Sg f TRUTH	O ho G3739 pr Acc Sg n WHICH	O ho G3588 t_ Nom Sg m THE	ΚΟΣΜΟΣ kosmos G2889 n_ Nom Sg m SYSTEM world	ΟΥ ou G3756 Part Neg NOT	[17] [Even] the Spirit of truth; whom the world cannot receive, because it seeth him not, neither knoweth him: but ye know him; for he dwelleth with you, and shall be in you.
	ΔΥΝΑΤΑΙ dunatai G1410 vi Pres midD/pasD 3 Sg IS-ABLE can	ΛΑΒΕΙΝ labein G2983 vn 2Aor Act TO-BE-GETTING	ΟΤΙ hoti G3754 Conj that	ΟΥ ou G3756 Part Neg NOT	ΘΕΩΡΕΙ theOrei G2334 vi Pres Act 3 Sg it-IS-beholdING	ΑΥΤΟ auto G846 pp Acc Sg n it	ΟΥΔΕ oude G3761 Adv NOT-YET neither	ΓΙΝΩΣΚΕΙ ginOskei G1097 vi Pres Act 3 Sg IS-KNOWING	ΑΥΤΟ auto G846 pp Acc Sg n it	
	ΥΜΕΙΣ humeis G5210 pp 2 Nom Pl YOU(p) ye	ΔΕ de G1161 Conj YET	ΓΙΝΩΣΚΕΤΕ ginOskete G1097 vi Pres Act 2 Pl ARE-KNOWING	ΑΥΤΟ auto G846 pp Acc Sg n it	ΟΤΙ hoti G3754 Conj that	ΠΑΡ par G3844 Prep BESIDE	ΥΜΙΝ humin G5213 pp 2 Dat Pl to-YOU(p) ye	ΜΕΝΕΙ menei G3306 vi Pres Act 3 Sg it-IS-REMAINING	ΚΑΙ kai G2532 Conj AND	ΕΝ en G1722 Prep IN	ΥΜΙΝ humin G5213 pp 2 Dat Pl YOU(p) ye
	ΕΣΤΑΙ estai G2071 vi Fut vxx 3 Sg SHALL-BE									

Figure 29. Spirit is 'IT' source ISA

Note: the above Interlinear translation shows that the spirit is an "it" and is gender neuter! There is no "him" or "he" intended in the original word of God, and it is, in my opinion, disingenuous to change what God intended to fit the deceiver's agenda.

Here's a more correct translation of John 14:17 from the New World Translation:

> 17 the spirit of the truth, + which the world cannot receive, because it neither sees it nor knows it. + You know it, because it remains with you and is in you. 18 I will not leave you bereaved. * I am coming to you. +

Figure 30. Proper Gender Usage, as per JW Bible

Also, note that it is the spiritual Christ and Father (they are the spiritual oneness) who are coming to us, and it is Christ and the Father who are the spiritual Comforter within us, which makes us Christian. If you don't have Christ in you, then you are none of his. And it is the Father in spiritual form within us who will raise us up, just as he raised Christ. The Father and the Son are the Spirit that dwells in us, and it is the Spirit and the bride that go forth into the universe.

> *You, however, are controlled not by the sinful nature but by the Spirit, if the Spirit of God lives in you. **And if anyone does not have the Spirit of Christ, he does not belong to Christ.** But if Christ is in you, your body is dead because of sin, yet your spirit is alive because of righteousness. **And if the Spirit of him who raised Jesus from the dead is living in you**, he who raised Christ from the dead will also give life to your mortal bodies through his Spirit, who lives in you. (Romans 8:9–11, NIV)*

And in the New World Translation of the same passage:

> 9 However, you are in harmony, not with the flesh, but with the spirit, + if God's spirit truly dwells in you. But if anyone does not have Christ's spirit, this person does not belong to him. 10 But if Christ is in union with you, + the body is dead because of sin, but the spirit is life because of righteousness. 11 If, now, the spirit of him who raised up Jesus from the dead dwells in you, the one who raised up Christ Jesus from the dead + will also make your mortal bodies alive + through his spirit that resides in you.

Figure 31. Proper Gender Usage, as per JW Bible

It is God's Spirit essence that lives in us, and the personalities that emit their oneness in form of a liquid spiritual river are the Father and Son Jesus.

The Comforter is therefore the combined presence of the Father and the Son Jesus. They comfort us by being present in us by the extension of their living water like essence they pour into us. The source is the Father and the Son, who are the spirit, not some other third person.

From one of the most famous prophecies in the book of Isaiah, shown in a new light, because Christ and the Father are one, and together they are the promised Counselor in us:

> *For a **Child hath been born to us**, A Son hath been given to us, And the princely power is on his shoulder, And He doth call his name **Wonderful**, **Counsellor**, **Mighty God**, **Father of Eternity**, **Prince of Peace**. To the increase of the princely power, And of peace, there is no end, On the throne of David, and on his kingdom, To establish it, and to support it, In judgment and in righteousness, Henceforth, even unto the age, The zeal of Jehovah of Hosts doth this. (Isaiah 9:6–7, YLT)*
>
> *I and the Father are one. (John 10:30, YLT)*

The Comforter is not some third spirit person; it is the combined spiritual presence of Jesus and the Father. There are no other personalities talked about in the previous few scriptures other than Christ and the Father, who raised Christ from the dead. What an embarrassment it would be to exclude some supposed equal third person of the Godhead!

The doctrine of the Trinity states that the Holy Ghost is a separate third person of the Godhead. I believe the spirit who invented the Trinity is a brilliant imposter. This imposter spirit has deceived most of the world into believing that he is Father God. How? Let me explain.

I ask you, which person in the Godhead is the Father of Jesus? But please read this before you answer:

> *"How will this be," Mary asked the angel, "since I am a virgin?"*
> *The angel answered, "**The Holy Spirit will come upon you, and the power of the Most High will overshadow you**. So the holy one to be born will be called the Son of God. (Luke 1:34–35, NIV)*
>
> *This is how the birth of Jesus Christ came about: His mother Mary was pledged to be married to Joseph, but before they came together, she was found to **be with child through the Holy Spirit**. Because Joseph her husband was a righteous man and did not want to expose her to public disgrace, he had in mind to divorce her quietly. But after he had considered this, an angel of the Lord appeared to him in a dream and said, "Joseph son of David, do not be afraid to take Mary home as your wife, because **what is conceived in her is from the Holy Spirit**. She will give birth to a son, and you are to give him the name Jesus, because he will save his people from their sins." (Matthew 1:18–21, NIV)*

So now, after reading the above scriptures, please answer, who is the Father of Jesus?

I say, if you believe there are three persons in the Godhead, and if the Holy Ghost is the third person, then the Holy Ghost is the father! That is his lie all along. Remember when Satan told Jesus he would give him all the kingdoms, if he would just "worship him" (Satan was implying or inserting himself as God the Father). Christ replied, "Get behind me, Satan! Only God is worthy of worship!"

Satan is an imposter, pawning himself off as God. And I say inserting himself, or overlaying himself, in place of the Holy Ghost.

(There is a third spirit personality walking about the throne of God, and his name is Satan- the antichrist.

> *Don't let anyone deceive you in any way, for [that day will not come] until the rebellion occurs and the man of lawlessness is revealed, the man doomed to destruction. [4]* **He will oppose and will exalt himself over everything that is called God or is worshiped, so that he sets himself up in God's temple, proclaiming himself to be God.** *2 Thessalonians 2:3 (NIV)*

Now this is the truth from God the Father person: *John 1:14 (NIV):* "**The Word became flesh** *and made his dwelling among us. We have seen his glory, the glory of the One and Only,* **who came from the Father***, full of grace and truth."*

The Holy Ghost is the spirit unity, or essence, of God the Father and Jesus the Son. The Lord is the Spirit (2 Corinthians 3:17), and the Spirit is the powerful living water extension of the most high (Luke 1:35).

God is made of spirit (John 4:24), and so shall we later be (1 Corinthians 15:50–52), as part of an expanded spiritual unity of God (John 17:20–22). The Spirit, or Holy Ghost, is not a separate third person—it is the essence of the Father, the Son, and in their likeness, eventually the bride.

Even the very words of God are his spiritual essence and life (not a third person)!

The Spirit gives life; the flesh counts for nothing. ***The words** I have spoken to you **are spirit and they are life**. [64] Yet there are some of you who do not believe." For Jesus had known from the beginning which of them did not believe and who would betray him. (John 6:63–64, NIV)*

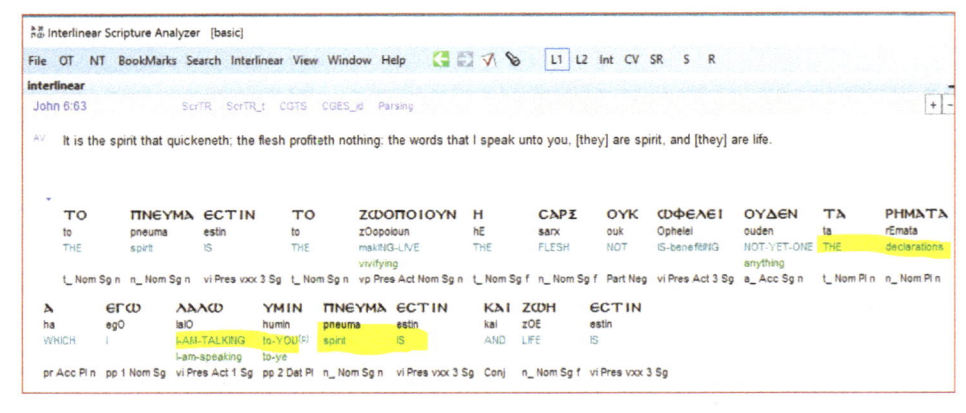

Figure 32. The Words of God, are Spirit (ISA)

Another question for Trinitarians I would also ask, who raised Jesus from the dead? This is an especially important question since we just read from Romans 8:9–11 that if the spirit of him who raised Jesus dwells in us, he will also raise us. But please read this before you answer.

*I keep asking that the God of our Lord Jesus Christ, the **glorious Father, may give you the Spirit of wisdom and revelation, so that you may know him better.** I pray also that the eyes of your heart may be enlightened in order that you may know the hope to which he has called you, the riches of his glorious inheritance in the saints, **and his incomparably great power for us who believe. That power is like the working of his mighty strength, which he exerted in Christ when***

> ***he raised him from the dead*** *and seated him at his right hand in the heavenly realms, far above all rule and authority, power and dominion, and every title that can be given, not only in the present age but also in the one to come. And God placed all things under his feet and appointed him to be head over everything for the church, which is his body, the fullness of him who fills everything in every way. (Ephesians 1:17–23, NIV)*

> *"To you first, God, having raised up His child Jesus, did send him, blessing you, in the turning away of each one from your evil ways." (Acts 3:26, YLT)*

> *Paul, an apostle—not from men, nor through man, but through Jesus Christ, and God the Father, who did raise him out of the dead. (Galatians 1:1, YLT)*

God the Father raised Jesus by exerting His powerful (liquid-like) spirit. If the Holy Spirit was a third person in the God Head, then it was the third person, or Holy Ghost that raised Jesus. Not so, the Father raised Jesus by His spiritual will.

The spirit is the essence of Christ and the Father, later to include us in a spiritual oneness, with us in their likeness as the bride of Christ! And we, then combined as one, head out to include any who are thirsty and wish to partake of God's oneness (family).

And this gets back to *Revelation 22:17: "The Spirit and the bride say, 'Come!' And let him who hears say, 'Come!' Whoever is thirsty, let him come; and whoever wishes, let him take the free gift of the water of life" (NIV)*.

I mentioned there are hundreds of these mistranslations in the New Testament, where the male gender is flaunted on the neuter gender. I have found that a lot of people know that, and accept it as okay for the following reason.

Firstly, it seems to me they want to believe it (why, I don't know), and secondly there is shred of excuse in the following scripture. In John (14, 15 and 16), and only in this gospel of the New Testament, the male gender does correctly apply! It is where the word *comforter* is applied to the spirit. But this is a personification and not directly the spirit.

For example, you would never say to a person the following point about death: "**Death, he** will get you in the end." That would be improperly applying the male gender to the neuter gender "death." However, if you personify death, by using the name Grim Reaper, you could then use the male gender correctly. "The **Grim Reaper, he** will get you in the end." That is why it is correct to use the male gender when the personification "comforter" is applied to the neuter "spirit."

The Spirit of God is an "it" and not a "he."[22]

Below is John 14:26 as shown in the Interlinear Scripture Analyzer software.

Note that the third Greek word shown, translate to "consoler" in green, has the male gender as discussed above ("n_Nom Sg m" has the male gender). Also note that just two words later, the word "spirit," is shown as neuter ("n Nom Sg n" has the neuter gender as usual).

Figure 33. The Comforter-Personification has Male Gender (ISA)

```
AV  But the Comforter, [which is] the Holy Ghost, whom the Father will send in my name, he shall teach you all things, and bring all things to your
    remembrance, whatsoever I have said unto you.

O           ΔΕ      ΠΑΡΑΚΛΗΤΟΣ     ΤΟ      ΠΝΕΥΜΑ      ΤΟ      ΑΓΙΟΝ      Ο       ΠΕΜΨΕΙ              Ο       ΠΑΤΗΡ       ΕΝ
ho          de      paraklEtos     to      pneuma      to      hagion     ho      pempsei             ho      patEr        en
THE         YET     BESIDE-CALLer  THE     spirit      THE     HOLY       WHICH   SHALL-BE-SENDING    THE     FATHER       IN
                    consoler
t_Nom Sg m  Conj    n_Nom Sg m     t_Nom Sg n  n_Nom Sg n  t_Nom Sg n  a_Nom Sg n  pr Acc Sg n  vi Fut Act 3 Sg   t_Nom Sg n  n_Nom Sg m  Prep

ΤΩ          ΟΝΟΜΑΤΙ     ΜΟΥ     ΕΚΕΙΝΟΣ     ΥΜΑΣ        ΔΙΔΑΞΕΙ             ΠΑΝΤΑ   ΚΑΙ     ΥΠΟΜΝΗΣΕΙ               ΥΜΑΣ        ΠΑΝΤΑ
tO          onomati     mou     ekeinos     humas       didaxei             panta   kai     hupomnEsei              humas       panta
THE         NAME        OF-ME   that        YOU(P)      SHALL-BE-TEACHING   ALL     AND     SHALL-BE-UNDER-REMINDING YOU(P)     ALL
                                that-one    ye                                              shall-be-reminding       ye
t_Dat Sg n  n_Dat Sg n  pp 1 Gen Sg  pd Nom Sg m  pp 2 Acc Pl  vi Fut Act 3 Sg  a_Acc Pl n  Conj  vi Fut Act 3 Sg    pp 2 Acc Pl  a_Acc Pl n

Α           ΕΙΠΟΝ       ΥΜΙΝ
ha          eipon       humin
WHICH       I-said      to-YOU(P)
                        to-ye
pr Acc Pl n  vi 2Aor Act 1 Sg  pp 2 Dat Pl
```

I have expanded a part of the scripture below to make it easier to see, and I have included the definitions for the parsing in the notes section.

ΠΑΡΑΚΛΗΤΟC	ΤΟ	ΠΝΕΥΜΑ
paraklEtos	to	pneuma
BESIDE-CALLer	THE	spirit
consoler		
n_ Nom Sg m	t_ Nom Sg n	n_ Nom Sg n

Figure 34. The Comforter, Consoler, BESIDE-CALLer, has Male Gender (ISA)

That's not IT—there is more.

An entire verse was inserted into the Bible, so that the Trinity doctrine could apparently have scriptural foundation. That is amazingly bold and dangerous for the translators who stooped to that level, especially considering the warning given in the Bible:

> *I warn everyone who hears the words of the prophecy of this book: If anyone adds anything to them, God will add to him the plagues described in this book. And if anyone takes words away from this book of prophecy, God will take away from him his share in the tree of life and in the holy city, which are described in this book. (Revelation 22:18–19, NIV)*

> *"Every word of God is flawless;*
> *he is a shield to those who take refuge in him.*
> *Do not add to his words,*
> *or he will rebuke you and prove you a liar. (Proverbs 30:5–6, NIV)*

The words of man, shown below 1 John 5:8, were inserted into the Latin Vulgate in the sixteenth century, with the agenda to add to the word of God a basis for the false doctrine of the Trinity. Shame.

First John 5:6–12 (NIV):

> **6** This is the one who came by water and blood*—Jesus Christ. He did not come by water only, but by water and blood. And it is the Spirit who testifies, because the Spirit is the truth.* **7** For there are three* that testify: **8** the*[Late manuscripts of the Vulgate *testify in heaven: the Father, the Word and the Holy Spirit, and these three are one. 8And there are three that testify on earth: the* (not found in any Greek manuscript before the sixteenth century)] Spirit, the water and the blood; and the three are in agreement. **9** We accept man's testimony,* but God's testimony is greater because it is the testimony of God,* which he has given about his Son. **10** Anyone who believes in the Son of God has this testimony in his heart.* Anyone who does not believe God has made him out to be a liar,* because he has not believed the testimony God has given about his Son. **11** And this is the testimony: God has given us eternal life,* and this life is in his Son.* **12** He who has the Son has life; he who does not have the Son of God does not have life.*

Figure 35. Man made—false scripture, source Laridian Pocket Bible

So we see that the words **"the Father, the Word and the Holy Spirit, and these three are one. And there are three that testify on earth"** were irresponsibly added by man with the intent to include the Trinity in the word of God, the Bible. Again, shame!

Hopefully, I have revealed to you the agenda to change the Bible by inserting a view of a spirit person as the third member of the Godhead. Of mistranslating scripture because it gives male gender to that person (that is not a person, but is the essence of what God is made of).

I would like to finish off this chapter now by looking at an apparent contradiction to all I have been saying. It is a scripture that is mistakenly used to prove there are three persons in a Trinity of the Godhead. Actually, there are two main scriptures that Trinitarians use.

First: *Matthew 28:19 (YLT): "Having gone, then, disciple all the nations, (baptizing them--to the name of the Father, and of the Son, and of the Holy Spirit."*

WHAT IS THE NAME WE ARE BAPTIZED INTO?

From a webpage called Netbiblestudy.net, I would like to print from one of their studies on being baptized into Christ.[23]

SPACE THE TRUE FRONTIER!

> **Baptized Into Christ**
>
> Notice that it is baptism that puts us **into Christ** so we can be saved. Galatians 3:27, "For as many of you as were **baptized into Christ have put on Christ.**" Baptism is the only way a person can get **into Christ**. The preposition *into* indicates a change of relationship. You can search from the front to the back of your Bible, and you will find no other way to get **into Christ** except by being baptized into Christ. Also Romans 6:3 says the same thing, "Or do you not know, that as many of us as **were baptized into Christ Jesus** were baptized into his death?" These are the only two passages in the entire New Testament that tell us how we get into Christ. So the only way we can get **into Christ** is to be **baptized into Christ**. Either a person has been **baptized into Christ,** or he is still outside of Christ. We are either in or we are out. Our hope is based on our being in Christ. Have you been **baptized into Christ** in order to be saved and have your sins forgiven? If you haven't been, let us notice some of the things that you are missing.
>
> 1) Ephesians 1:3 says, "Blessed be the God and Father of our Lord Jesus Christ, who has blessed us with **every spiritual blessing** in heavenly places **in Christ.**" This verse tells us that every spiritual blessing is located **in Christ**. Only those who are **in Christ** can enjoy these spiritual blessings. Since this verse says every spiritual blessing is in Christ, then there can be no spiritual blessings for those who are outside of Christ. If you have not been baptized into Christ, you are still outside of Christ, and you are not entitled to any spiritual blessings. The Lord arranged it that way and no man has the right to change it.

Figure 36. Into the Name of Christ (Lesson_17_oes http://www.netbiblestudy.net/new_page_17.htm)

So we baptized into Christ, and at the same time "sealed" in the Holy Spirit:

> *In order that we, who were the first to hope in Christ, might be for the praise of his glory. And you also were included in Christ when you heard the word of truth, the gospel of your salvation. Having believed, you were marked in* **him with a seal, the promised Holy Spirit***, who is a deposit guaranteeing our inheritance until the redemption of those who are God's possession—to the praise of his glory. (Ephesians 1:12–14, NIV)*

As Jesus was given the Father's name, so are we. *John 17:11 (NIV):* "*I will remain in the world no longer, but they are still in the world, and I am coming to you.* **Holy Father, protect them by the power of your name—the name you gave me**—*so that they may be one as we are one.*"

I would like to comment that God has given us a ubiquitous example of fathers giving their children their name. When we were born as physical babies, we were given our father's name. The same is true of our spiritual family in God. And as we were enveloped in the water of the womb of our mothers, so are we now enveloped (sealed) into the living waters of our down payment before our complete change into the likeness of God—spiritually at one with Jesus and the Father, given the Father's name, which is the name he gave Jesus, and sealed into the spirit essence of the family of God (1 Corinthians 15:42–49, 50–53).

We are baptized into Christ, which gives us access to oneness with the Father, and we are sealed in the life essence of the living water, that is the spirit. The name is the Fathers! There are not three different names we are baptized into.

> *Then I looked, and there before me was the Lamb, standing on Mount Zion, and with him 144,000 **who had his name and his Father's name written on their foreheads**. (Revelation 14:1, NIV)*

Consider why is the name of the "holy spirit" not mentioned here? Because the Holy Spirit is not a separate person—the Father and Jesus are the combined Holy Spirit essence!

We are not baptized into the name of a third person spirit. We are sealed in the living water of the womb of our transition into the oneness of God. Christ, of course, has some unique names, but when it comes to the unity of God, he has the Father's name, and that is the name we also get. I repeat John 17:11: *I will remain in the world no longer, but they are still in the world, and I am coming to you.* **Holy Father, protect them by the power of your name—the name you gave me**—*so that they may be one as we are one.*

When we come into, are born into, and baptized into the family of God, it has the name of the Father, the same name he gave Christ, sealed in the body of living water spirit essence of God.

> *"Righteous Father, though the world does not know you, I know you, and they know that you have sent me. [26] I have made you known to them, and will continue to make you known in order that the love you have for me may be in them and that I myself may be in them." (John 17:25–26, NIV)*

Christ is the great I Am, or Logos Lord of the Old Testament. He is the one who made everything. That is why he had to come and reveal the Father. We had never heard or seen the Father.

> ***Jesus said*** *unto them, Verily, verily, I say unto you, Before Abraham was,* ***I am.*** *(John 8:58, KJV)*

> *In the beginning was the Word, and the Word was with God, and the Word was God. He was with God in the beginning.* ***Through him all things were made;*** *without him nothing was made that has been made. (John 1:1–3, NIV)*

And how many beings have the name of God?

> *Who hath wrought and done it, calling the generations from the beginning? I the LORD, the first, and with the last; I am he. (Isaiah 41:4, KJV)*

> *Thus saith the LORD the King of Israel, and his redeemer the LORD of hosts; I am the first, and I am the last; and beside me there is no God. (Isaiah 44:6, KJV)*

> *I am Alpha and Omega, the beginning and the ending, saith the Lord, which is, and which was, and which is to come, the Almighty.*
> *I John, who also am your brother, and companion in tribulation, and in the kingdom and patience of Jesus Christ, was in the isle that is called Patmos, for the word of God, and for the testimony of Jesus Christ.*
> *I was in the Spirit on the Lord's day, and heard behind me a great voice, as of a trumpet,*
> *Saying, I am Alpha and Omega, the first and the last: and, What thou seest, write in a book, and send it unto the seven churches which are in Asia; unto Ephesus, and unto Smyrna, and unto Pergamos, and unto Thyatira,*

and unto Sardis, and unto Philadelphia, and unto Laodicea. (Revelation 1:8–11, KJV)

I Am that I Am

From Wikipedia, the free encyclopedia

I Am that I Am (אֶהְיֶה אֲשֶׁר אֶהְיֶה, *ehyeh ašer ehyeh* [ehˈje aˈʃer ehˈje]) is the common English translation (JPS among others) of the response that God used in the Hebrew Bible when Moses asked for his name (Exodus 3:14). It is one of the most famous verses in the Torah. *Hayah* means "existed" in Hebrew; *ehyeh* is the first person singular imperfect form and is usually translated in English Bibles as "I am" or "I will be" (or "I shall be"), for example, at Exodus 3:14. *Ehyeh asher ehyeh* literally translates as "I Am Who I Am." The ancient Hebrew of Exodus 3:14 lacks a future tense such as modern English has, yet a few translations render this name as "I Will Be What I Will Be," given the context of Yahweh's promising to be with his people through their future troubles.[1] Both the literal present tense "I Am" and the future tense "I will be" have given rise to many attendant theological and mystical implications in Jewish tradition. However, in most English Bibles, in particular the King James Version, the phrase is rendered as *I am that I am*.

Figure 37. The Name of God (Wikipedia 'I Am that I Am')

The final such claim in the Old Testament is found in Malachi 3:6, in which God appropriately reminds everyone that *"I am the Lord, I change not."* He is the great "I AM," the self-existent God. We must remember always that our own personal Savior and Lord Jesus Christ has revealed to us that He Himself is that same great I AM.

It is in the Gospel of John, however, that the most beautiful and personally meaningful "I am"s are found. There are seven of these, as follows:

"I am the bread of life" (John 6:35,48,51).

"I am the light of the world" (John 8:12).

"I am the door of the sheep"(John 10:7,9).

"I am the good shepherd" (John 10:11,14).

"I am the resurrection, and the life" (John 11:25).

"I am the way, the truth, and the life" (John 14:6).

"I am the true vine" (John 15:1,5).

How anyone could hear or see these claims and then still deny that they were claims to deity is a great mystery. For example, how could anyone except God Himself claim to be the way, the truth, and the life?

Figure 38. The Great I Am's (http://www.icr.org/article/i-ams-christ/)

One of the starkest examples of Christ being the "I Am":

Jesus, therefore, knowing all things that are coming upon him, having gone forth, said to them, "Whom do ye seek?" they answered him, "Jesus the Nazarene;" Jesus saith to them, ==*"I am* he;"== *—and Judas who delivered him up was standing with them;—when, therefore, he said to them—*=="I am *he,"*== ==*they went away backward, and fell to the ground.*== *Again, therefore, he questioned them, "Whom do ye seek?" and they said, "Jesus the Nazarene." (John 18:4–7, YLT)*

And from the Interlinear Scripture Analyzer, showing that when Christ as God, said, "I Am," they fell back onto the ground because he spoke his powerful God name.

Figure 39. The Great I Am, and they fell back (ISA)

We are baptized into Christ, which simultaneously gets us into oneness with the Father, in a spiritual essence that is the great "I Am." The Holy Spirit is what the "I Am" is made of. The "I Am" is Christ.

I would also like to sum up the point I am making in light of Matthew 28:19. We are baptized into the family of God, which has the Father's name, which has the Son, and is made of the spirit living water that seals us in unity. Just like our human family, where we get the father's family name, the same should be true in the family of God. We gain our Father's name in a oneness with the Son in the living essence of the spirit.

We are not baptized into three different persons' names; we are baptized into the oneness (family) of God.

Figure 40. Into the name of the Father, which He Gave the Son, as First, in the Spirit Sealed Family Entity (God) (ISA)

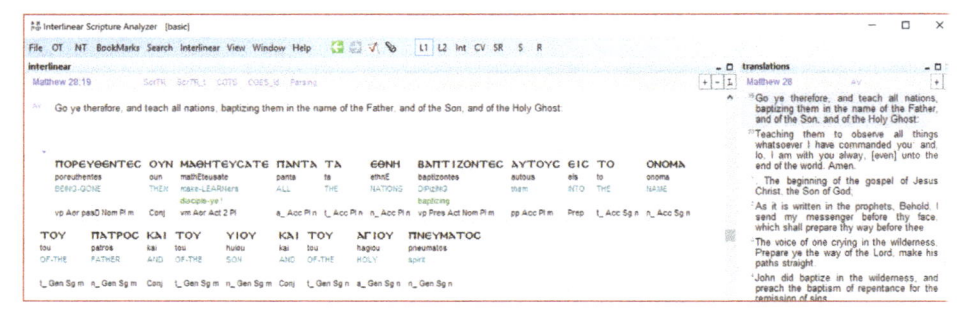

Second:

The grace of the Lord Jesus Christ, and the love of God, and the fellowship of the Holy Spirit, is with you all! Amen. (2 Corinthians 13:14, YLT)

In the above passage, there is reference to Jesus Christ, God, and the Holy Spirit. So how could it be possible to have a separate scripture that excludes the Holy Spirit yet is talking about the same subject of our fellowship?

That which was from the beginning, which we have heard, which we have seen with our eyes, which we have looked upon, and our hands have handled, of the Word of life;
(For the life was manifested, and we have seen it, and bear witness, and shew unto you that eternal life, which was with the Father, and was manifested unto us;)
That which we have seen and heard declare we unto you, that ye also may have fellowship with us: **and truly our fellowship is with the Father, and with his Son Jesus Christ.**

*And these things write we unto you, **that your joy may be full**. (1 John 1:1–4, KJV)*

This verse shows that our joy is complete in the unity with Father and Son. So where is the Holy Spirit person?

But of course, there is a Holy Spirit that is them (it is what they are made of). The Holy Spirit is a part of our joy too. Just not as a person. That is why in some verses the Holy Spirit is not talked about, or shown to be personally there, but must be there. The Holy Spirit is always there, wherever God is; in fact, God is spirit (John 4:24, 2 Corinthians 3:17–18).

In chapter 5 ("Knock, Knock"), I will show many of examples where the Holy Spirit is seemingly left out when discussing the Godhead. But the spirit of God is always there by composition. Only the Father and Son are seemingly present, but they are spirit essence—they are Holy Spirit (John 4:24, 2 Corinthians 3:17).

CHAPTER 3
HICCUPS ALONG THE ROAD TO THE UNIVERSE

The BATTLE PLANS OF THE PRINCE OF DARKNESS CRUSHED (but it comes with bruises)

> *So the LORD God said to the serpent, "Because you have done this,*
> *"Cursed are you above all the livestock*
> *and all the wild animals!*
> *You will crawl on your belly*
> *and you will eat dust*
> *all the days of your life.*
> *And I will put enmity*
> *between you and the woman,*
> *and between your offspring and hers;*
> *he will **crush your head**,*
> *and you **will strike his heel**." (Genesis 3:14–15, NIV)*

The so-called fall of man was not a surprise. Not a point for a do-over to God (or any kind of failure of God) in man's inabilities and our being too weak to deal with Satan. Up until Christ, man did not stay true to God and worshipped Satan by bending to Satan's evil will.

God knew from before the creation of the world this would happen. *Revelation 13:8 (KJV): "And all that dwell upon the earth shall worship him, whose names are not written in the book of life of the Lamb slain from the foundation of the world."* That's right. God knew man would fail in resisting the devil. Satan is the master prince of power of the air, and he knows us and our failings all too well. Satan was (I speculate) totally certain that **no man** could ever stand up against him. So by preventing man from staying true to God, Satan could make God's will fail, and therefore, he would be above God, by defeating God's will.

I think Satan deceived himself into truly believing that! To him, this was such a sure thing; he was convinced that he could audaciously stand up to the creator, and rebel. He knew there was no way that God's will could be accomplished through a human man, against Satan's enmity. And Satan was almost right, wasn't he?

In fact, I speculate that this was God's ace card and that it was planned all along as the ultimate incontestable way to keep evil out of eternity. The purpose being that the outcome of this calamity would ultimately show God who would forever remain loyal, showing God who should be given access to the tree of life and live forever, as part of God's family.

From this precedent of some of the created angels and the chief of angels rebelling, it would also be clear to us that we could never remain true to God on our own, no matter how righteous or how great, we might think we are. It takes Christ and the Father in us, and we must choose their presence with all our heart.

Satan, in his haughtiness, could not conceive of this level of humility that God would display in his love for us. That was Satan's downfall. I believe it was inconceivable to Satan that a God would possibly humble Himself so much. For man to accomplish the Fathers will, it took the 'Last Adam', man begotten of God.

God had to become man for man to be able to overcome Satan, and Satan never saw it coming—THE GOD WHO WOULD BE MAN.

We believers have been sharing a journey through time and space with all humanity since creation (or, I speculate, since the re-creation in Genesis). So where have we gotten in our travels over that time? Are we closer to inheriting that awesome frontier of space? I think most would agree mankind is in trouble, and if something doesn't happen soon, we could even become extinct.

What, or who, can possibly save us?

Many people believe Christ was a great prophet but do not accept his deity. Muslims and many Christians included. I think only

God, and no prophet can save us. And we are told Jesus is our Savior. *Acts 4:12(NIV):* "*Salvation is found in no one else, for* **there is no other name** *under heaven given to mankind* **by which we must be saved.**"

Jesus is our Savior God. No man can save us—only Jesus the man-God.

In my life, there had always been two possible sources of truth (which I faithfully considered trustworthy and credible)—the Scientists (with evolution) or the Bible and God. Early on, I was taught and believed evolution; it was simply a fact in elementary schools in the sixties. I grew strong in that faith, which proved extremely difficult to overcome intellectually.

It took miracles, but I have shifted from believing science, with evolution and all man's great accomplishments, to believing only God has the ultimate answers. Why and how did we get here? Despite all the abilities God has given man and all the good things accomplished, we have generally rejected God and find evil prevailing.

As a species, we still think we can survive on our own, and that nothing planned for us, we just happened by chance. We even go so far as to think we can obtain access to space without God. And both science and Christians have been blind to the fact that space is the place where God has been preparing us for all along. To me, it is like the true space race was not between Russia and the USA but between science and God. So will it be God or man that gets us there?

I remember President Kennedy's bold and wonderful speech, where he dedicated the United States to landing on the moon by the end of the decade. Even as a young child, now that was brave and wonderful. I pretty much spent the whole week watching television in July 1969 when it became one of the greatest realities mankind has ever accomplished! I remember so well all the triumphs and tragedies along the way. And I have tears in my eyes and pride in my heart at this very writing. We are wonderfully made, and it seems we can accomplish anything when we put our minds to it.

Some of the greatest thinking men *(Genesis 1:27 (KJV): "So God created man in his own image, in the image of God created he him;* **male and female** *created he them")* of all times see no need for a creator and, in fact, seem to hate the idea. But they do see the mess we are all in, and are greatly troubled by it. I no longer have faith in science to save us, and I don't believe evolution is our creator. Science, to me, is like a religion, that refuses to be questioned (their theories are facts). They are in competition with God, having a space-race to achieve the venue or garden he has prepared for us without him and on our own.

Sometimes, in the extreme of science as a religion, it is as if they are competing for our faith and worship, just like Satan competes with God. That may explain their sometimes, absolute hateful disdain for believers. They are dogmatic and believe we evolved from nothing (very slowly). Science seems tied to the idea that only observable physical realities exist and the idea of a spiritual reality is not only unobservable but hated nonsense.

I initially believed the science-religion wholeheartedly. Now thankfully, and contradictory to the THEORY of evolution, I see a life more realistically, where complex systems, on their own, become less complex and not more complex as evolution would require (things wind down and not up). This is the second law of thermodynamics and is called entropy. It is a fact that things are headed towards absolute zero, motionlessness. You can call it the big wind-down, and the universe is headed there.

There are isolated temporary exceptions to entropy, wherever some intelligent force does something to the contrary, like someone picking up random pool balls and stacking them. This takes energy and thought, but on its own, the universe is currently like a clock winding down, and there will have to be a new heaven created for our promised eternity to prevail.

Scientists discovered entropy, which God ordained, but they refuse to see that some intelligent force initially made the purposeful order of our (or wound-up) universe. They say our organized universe happened by chance, without intelligent will (force), just some dumb Higgs boson field–like thing. In the face of infinite observations (the very foundational premise of scientific method), they overlook and faithfully accept that blind ignorant chance forces made the most complex incredible creation that can hardly even be fully observed or understood! ==Because they will not admit God.==

Christ the Wisdom and Power of God
For the message of the cross is foolishness to those who are perishing, but to us who are being saved it is the power of God. For it is written:
"I will destroy the wisdom of the wise;
the intelligence of the intelligent I will frustrate."
*Where is the wise man? Where is the scholar? Where is the philosopher of this age****? Has not God made foolish the wisdom of the world? [For since in the wisdom of God*** **the world through its wisdom did not know him***, God was pleased through the foolishness of what was preached to save those who believe.*
(1 Corinthians 1:18–21, NIV)

To me, simple observation (like a scientific experiment) reveals that a pyramid stack of pool balls will decay into being randomly distributed throughout my car on a road trip to Banff. And it has never been observed that given enough time, they could by chance become a nice neat stack (perhaps on the trillionth trip?). **Surely common sense, with our eyes finally open, should tell us if energy and intelligence is required to stop entropy, then the big bang (wind-up-start) of it would require phenomenally more intelligence and energy.** For now, entropy prevails, and unless some intelligent force does something about it, the limits of the universe dictate that our hope for a nice neat stack of pool balls is doomed.

But science says that there is no God, no intelligent source force; it just needs a chance dumb force. Wow, now that takes faith. Do you believe that the given enough time a roomful of monkeys could type the *Encyclopedia Britannica*? Common sense tells me there are limits to this kind of limitless thinking, just like there are limits in calculus (the rabbit passes the turtle).

Recently, scientists have stated they have found the Higgs boson particle (or so-called God particle). They believe this proves the existence of the Higgs Boson field, which is special because this field is, to them, the source of all matter. If I understand science correctly, they believe it is a field that existed when there was only energy and no mass in the universe. Hence, they are delighted to say, in the beginning was the Higgs boson field and from that comes everything with mass (matter). They are happy because they have a (dumb non-intelligent) Higgs boson field as creator, replacing what they consider the foolish notion of God (intelligent creation).

This particle was mathematically predicted back in the 1960s to help balance out, or make correct, the new standard model of matter. Although, apparently, they are not satisfied with the outcome of their discovery, as I hear it is too light and should be much heavier and that maybe there are five different Higgs boson particles?

As I mentioned in the introduction to this book, I have a thought that since God is the creator, then perhaps everything that exists is sourced from Him, like frozen spirit (frozen Him). We Christians must accept what science has shown as the standard model and realize that our God made that model. I think that like there are many spectrums of light, some visible, some not, that (perhaps) when the scripture says "God is light," it is also literally true. At any rate, we must agree with what science is showing us and only differ on the source of this reality.

They say it came from nothing slowly. Christians say this is the miracle of how God did it, and I presume it simply really is Him. For if there was God alone before He created, then part of Him was the only available source for things—part of his existence in a different state, a state of matter. So the standard model may really be science looking at the depths of God, and in fact all be God particles.

Remember how the spirit of God vibrated like a wave at the Genesis creation and how God walks about the circle of the earth (quite insightful to have come from these ancient writings, don't you agree)? It took millennia for man and science to back up the word of God on these realities, if my guess of their meaning is correct.

*In the beginning of God's preparing the heavens and the earth-- [2] the earth hath existed waste and void, and darkness is on the face of the deep, and the **Spirit of God fluttering** on the face of the waters. (Genesis 1:1–2, YLT)*

Is not God high in heaven? And see the summit of the stars, That they are high. And thou hast said, "What-hath God known? Through thickness doth He judge? Thick clouds are a secret place to Him, And He doth not see;" ==**And the circle of the heavens He walketh habitually**==. *(Job 22:12–14, YLT)*

Notwithstanding, the competition between science and God, the signs and wonders of science are all around us. We all love the fruits of man's ingenuity. I just have a different premise to my faith versus their faith. **I also have a very different view of what is going to save us, not the reasonable good will of men, even if all were intelligent scientists.**

For me now, the creator God is the source and no longer a Higgs boson field and evolution (no matter how slowly). That aside, I want to contemplate something that I grew up with that is an outstanding accomplishment of scientific effort because I want to contrast this accomplishment next to God's, showing how it magnifies the greatness of God's achievement. This event and achievement has spanned my adult life. It makes a great statement about where we are now in the plan of God. Remember, the title of this book *Space the True Frontier,* and remember the true space race.

In September 5, 1977, the Voyager 1 space probe was launched by NASA to observe the outer reaches of our solar system. Since that time, Voyager has been travelling about 38,000 miles per hour away from earth. It is now the farthest artifact of man's presence in the universe, now almost 12.5 billion miles away (as of November 2015) and interstellar (outside our solar system).

No one expected Voyager to be so successful or to last so long. This has been going on in the background for most of my adult life and was even a part of an episode in *Star Trek*. It prompted a great scientist astronomer of our time, Carl Sagan, to a brilliant act. What would earth look like from way out there? It was not easy, but Carl Sagan, to his much-appreciated credit, was able to convince NASA to allow him to turn the camera in Voyager back around to take one last picture of earth.

This act was understandably resisted for good reason. There was a chance that the very dated equipment would be destroyed by such an action, and for what? Earth would look like almost nothing from that distance. It would be an insignificant dot, almost unfindable.

History has proven Carl Sagan correct, for it has amazed us all and takes our breath away. The most outstanding point is that the **insignificance of this blue dot is the significance**! And what an unbelievably great accomplishment of mankind it is. Even just getting the picture taken is astounding. Consider that light travels around the world over seven times in just one second. That's fast at 186,000 miles per second.

How long do you think it took for the radio waves to reach Voyager? Well, at the time the picture was taken, February 14, 1990, the space probe was about 3.75 billion miles away, and at 186,000 miles per second, that would be 5.6 hours. That is a 5.6 light hour time lag in the conversation. Amazing. But take the picture it did. It is called the blue dot and is phenomenal.

I have attached a link for you to check this out. Please do, as the words of Carl Sagan are unforgettable and true, and to hear him is worth it.

Mr. Sagan basically (and I think if more people were like him, it would certainly have a better chance of occurring) gives us one last chance to get our act together before we destroy the planet. He asks us to be reasonable and work together. I very much respect him and most of the intent of science in this regard. But I don't have faith man will do it. We should certainly try, but I really have come to see the depth of our inability to exist exclusive of God. We do not know how to handle the knowledge of good and evil and choose life successfully.

Carl Sagan's – Pale Blue Dot

****Add Your Comments in the Facebook Comments Section Below!*

> *From this distant vantage point, the Earth might not seem of any particular interest. But for us, it's different. Look again at that dot. That's here. That's home.*

That's us. On it everyone you love, everyone you know, everyone you ever heard of, every human being who ever was, lived out their lives. The aggregate of our joy and suffering, thousands of confident religions, ideologies, and economic doctrines, every hunter and forager, every hero and coward, every creator and destroyer of civilization, every king and peasant, every young couple in love, every mother and father, hopeful child, inventor and explorer, every teacher of morals, every corrupt politician, every "superstar," every "supreme leader," every saint and sinner in the history of our species lived there – on a mote of dust suspended in a sunbeam.

The Earth is a very small stage in a vast cosmic arena. Think of the rivers of blood spilled by all those generals and emperors, so that, in glory and triumph, they could become the momentary masters of a fraction of a dot. Think of the endless cruelties visited by the inhabitants of one corner of this pixel on the scarcely distinguishable inhabitants of some other corner, how frequent their misunderstandings, how eager they are to kill one another, how fervent their hatreds. Our posturings, our imagined self-importance, the delusion that we have some privileged position in the Universe, are challenged by this point of pale light.

Our planet is a lonely speck in the great enveloping cosmic dark. In our obscurity, in all this vastness, there is no hint that help will come from elsewhere to save us from ourselves. The Earth is the only world known so far to harbor life. There is nowhere else, at least in the near future, to which our species could migrate. Visit yes, Settle, not yet. Like it or not, for the moment the Earth is where we make our stand. It has been said that astronomy is a humbling and character building experience.

Figure 41. Carl Sagan, and the Famous Blue Dot (http://www.befreetoday.com.au/carl-sagan-pale-blue-dot)

There is perhaps no better demonstration of the folly of human conceits than this distant image of our tiny world. To me, it underscores our responsibility to deal more kindly with one another, and to preserve and cherish the pale blue dot, the only home we've ever known.

Figure 42. (Carl Sagan continued) (http://www.befreetoday.com.au/carl-sagan-pale-blue-dot/)

SPACE THE TRUE FRONTIER!

So if that is man's hope, what does God say?

Man still believes we have the ability to work things out, but Christ reveals the truth.

> ### *Signs of the End of the Age*
> *Jesus left the temple and was walking away when his disciples came up to him to call his attention to its buildings. "Do you see all these things?" he asked. "I tell you the truth, not one stone here will be left on another; every one will be thrown down."*
>
> *As Jesus was sitting on the Mount of Olives, the disciples came to him privately. "Tell us," they said, "when will this happen, and what will be the sign of your coming and of the end of the age?"*
>
> *Jesus answered: "Watch out that no one deceives you. For many will come in my name, claiming, 'I am the Christ,' and will deceive many. You will hear of wars and rumors of wars, but see to it that you are not alarmed. Such things must happen, but the end is still to come. Nation will rise against nation, and kingdom against kingdom. There will be famines and earthquakes in various places. All these are the beginning of birth pains.*
>
> *"Then you will be handed over to be persecuted and put to death, and you will be hated by all nations because of me. At that time many will turn away from the faith and will betray and hate each other, and many false prophets will appear and deceive many people. Because of the increase of wickedness, the love of most will grow cold, but he who stands firm to the end will be saved. And this gospel of the kingdom will be preached in the whole world as a testimony to all nations, and then the end will come.*
>
> *"So when you see standing in the holy place 'the abomination that causes desolation,' spoken of through the prophet Daniel—let the reader understand—then let those who are in Judea flee to the mountains. Let no one*

on the roof of his house go down to take anything out of the house. Let no one in the field go back to get his cloak. How dreadful it will be in those days for pregnant women and nursing mothers! Pray that your flight will not take place in winter or on the Sabbath. For then there will be great distress, unequaled from the beginning of the world until now—and never to be equaled again. ==**If those days had not been cut short, no one would survive**==*, but for the sake of the elect those days will be shortened. At that time if anyone says to you, 'Look, here is the Christ!' or, 'There he is!' do not believe it. For false Christs and false prophets will appear and perform great signs and miracles to deceive even the elect—if that were possible. See, I have told you ahead of time.*

"So if anyone tells you, 'There he is, out in the desert,' do not go out; or, 'Here he is, in the inner rooms,' do not believe it. For as lightning that comes from the east is visible even in the west, so will be the coming of the Son of Man. Wherever there is a carcass, there the vultures will gather.

"Immediately after the distress of those days

"'the sun will be darkened,
and the moon will not give its light;
the stars will fall from the sky,
and the heavenly bodies will be shaken.'

*"**At that time the sign of the Son of Man will appear in the sky,** and all the nations of the earth will mourn. They will see the Son of Man coming on the clouds of the sky, with power and great glory. **And he will send his angels with a loud trumpet call**, and they will gather his elect from the four winds, from one end of the heavens to the other.*

"Now learn this lesson from the fig tree: As soon as its twigs get tender and its leaves come out, you know that summer is near. Even so, when you see all these things, you know that it is near, right at the door. I tell you the truth, this generation will certainly not pass away until

all these things have happened. Heaven and earth will pass away, but my words will never pass away.

The Day and Hour Unknown
"No one knows about that day or hour, not even the angels in heaven, nor the Son, but only the Father. As it was in the days of Noah, so it will be at the coming of the Son of Man. For in the days before the flood, people were eating and drinking, marrying and giving in marriage, up to the day Noah entered the ark; and they knew nothing about what would happen until the flood came and took them all away. That is how it will be at the coming of the Son of Man. Two men will be in the field; one will be taken and the other left. Two women will be grinding with a hand mill; one will be taken and the other left.
"Therefore keep watch, because you do not know on what day your Lord will come. But understand this: If the owner of the house had known at what time of night the thief was coming, he would have kept watch and would not have let his house be broken into. So you also must be ready, because the Son of Man will come at an hour when you do not expect him.
"Who then is the faithful and wise servant, whom the master has put in charge of the servants in his household to give them their food at the proper time? It will be good for that servant whose master finds him doing so when he returns. I tell you the truth, he will put him in charge of all his possessions. But suppose that servant is wicked and says to himself, 'My master is staying away a long time,' and he then begins to beat his fellow servants and to eat and drink with drunkards. The master of that servant will come on a day when he does not expect him and at an hour he is not aware of. He will cut him to pieces and assign him a place with the hypocrites, where there will be weeping and gnashing of teeth. (Matthew 24:1–51, NIV)

Not man, not science, but the God who would be man, that's what it takes.

God says that without the return of Christ, mankind is doomed and that no flesh will be left alive (Matthew 24:22). Notice that it says, the world mourns at Christ's return. Why? Because it is deceived into believing a false god, the abomination that is at the holy place of God, the one acting the position of God, though he is not. Thank God that Christ is returning.

I want to close this chapter with one of the most heartfully appreciated realities of our Father and the depth of what He did for us through His only begotten Son.

In Matthew 24:36, we learn there are things that only the Father knows—things not even Christ knew. I speculate, perhaps one of those things came as an absolute shock to Christ, and the impossibility of it, was the demise of Satan.

Christ, as the man-God, was not alone but was always together with the Father. *John 16:32 (NIV): "But a time is coming, and has come, when you will be scattered, each to his own home. You will leave me all alone.* **Yet I am not alone, for my Father is with me."**

That is until the shock.

> From the sixth hour until the ninth hour darkness came over all the land. About the ninth hour Jesus cried out in a loud voice, *"Eloi, Eloi, lama sabachthani?"*—which means, *"My God, my God, why have you forsaken me?"* (Matthew 27:45–46, NIV)

Is it possible, that Jesus was 100 percent man at that moment, alone and forsaken from the Father? And is it possible, that He thereby accomplished what was surely impossible to Satan.

Man overcame Satan in Jesus on the stake—dying for us with bruised heels, nailed into wood at the very place of the skull, Golgotha, as He crushed the skull of Satan! *Matthew 27:33 (NIV): "They came to a place called Golgotha (which means The Place of the Skull)."*

Then the LORD God said to the woman, "What is this you have done?"
The woman said, "The serpent deceived me, and I ate."
So the LORD God said to the serpent, *"Because you have done this,*
"Cursed are you above all the livestock
and all the wild animals!
You will crawl on your belly
and you will eat dust
all the days of your life.
And I will put enmity
between you and the woman,
and between your offspring and hers;
he will crush your head,
and you will strike his heel." *(Genesis 3:13–15, NIV)*

Satan (with all his pride and vanity) **could not fathom the humility of the creator** of 13.7 billion light-years of "stuff'" coming to an insignificant blue dot as a poor carpenter—to accomplish the plan of God through His death.

For man to overcome Satan, God would have to become Man, and the only begotten Son, man-God Messiah, did. And to right all the wrong of sin (Satan's way), He had to die alone for all of us. That is an unimaginable sacrifice. **Praise to our Highest God, who only could have thought of it. Hallelujah and amen!**

Check out this video on YouTube, "GOD and the UNIVERSE," published by a channel called John 3:16.

CHAPTER 4: THE NOT FORGOTTEN

We are children of the promises, and are not forgotten. We are the spiritual temple of Israel!

<u>But we sure do forget</u>. Just like the example below from ancient Judah under King Josiah.

We have forgotten what the Bible says about the meaning and purpose of the temple. We forget that we are Israel, God's people, who eventually will rule with Him. And it seems we haven't heard about the glorious spacious firmament provided for our future spiritual eternity.

The New Covenant, talked about in the book of Hebrews, is not just for Jew's!

In fact the people of the Kingdom of Judah, and of Northern Israel, will need to be grafted back into the New Covenant olive branch, that is Christ!

Israel is now Christianity.

There is no New Testament for Jew's that is any different than the one for gentiles.

Hebrews 8:8 (NIV)

[8] But God found fault with the people and said:
"The time is coming, declares the Lord,
when I will make a new covenant
with the house of Israel
and with the house of Judah.
[9] It will not be like the covenant
I made with their forefathers
when I took them by the hand
to lead them out of Egypt,
because they did not remain faithful to my covenant,

and I turned away from them,
declares the Lord.
[10] This is the covenant I will make with the house of Israel
after that time, declares the Lord.
I will put my laws in their minds
and write them on their hearts.
I will be their God,
and they will be my people.
[11] No longer will a man teach his neighbor,
or a man his brother, saying, 'Know the Lord,'
because they will all know me,
from the least of them to the greatest.
[12] For I will forgive their wickedness
and will remember their sins no more."
[13] By calling this covenant "new," he has made the first one obsolete; and what is obsolete and aging will soon disappear.

Who is Peter talking to in First Peter 2:3? Is it only to Jews? Are they the only ones? Or is it to all believers? Yes of course.

Are we not the grafted in chosen people? Kings and Priests, in the coming Kingdom?

There are not two coming Kingdoms!

Jesus is the corner stone (or cap-stone) of the New Testament Temple of God. We, (because we have Christ in us), are part of that holy temple. We are living stones, are we not?

1 Peter 2:3 (NIV)
[3] now that you have tasted that the Lord is good.
The Living Stone and a Chosen People
[4] As you come to him, the living Stone—rejected by men but chosen by God and precious to him— [5] you also, like living stones, are being built into a spiritual house to be a holy priesthood, offering spiritual sacrifices acceptable to God through Jesus Christ. [6] For in Scripture it says:

"See, I lay a stone in Zion,
a chosen and precious cornerstone,
and the one who trusts in him
will never be put to shame."

[7] Now to you who believe, this stone is precious. But to those who do not believe,
"The stone the builders rejected
has become the capstone,"
[8] and,
"A stone that causes men to stumble
and a rock that makes them fall."
They stumble because they disobey the message—which is also what they were destined for.

[9] But you are a chosen people, a royal priesthood, a holy nation, a people belonging to God, that you may declare the praises of him who called you out of darkness into his wonderful light. [10] Once you were not a people, but now you are the people of God; once you had not received mercy, but now you have received mercy.

[11] Dear friends, I urge you, as aliens and strangers in the world, to abstain from sinful desires, which war against your soul. [12] Live such good lives among the pagans that, though they accuse you of doing wrong, they may see your good deeds and glorify God on the day he visits us.

Below is the crux of my observation. Israel is now a Christian thing, having believing Jew's, and grafted in gentile believers! In the future, when Christ has returned in His glorious state, He will rule together with kings and priests, that are living stones, that are the temple of God.

That is us!

Matthew 19:28 (NIV)

[28] Jesus said to them, "I tell you the truth, at the renewal of all things, when the Son of Man sits on his glorious throne, you who have followed me will also sit on twelve thrones, judging the twelve tribes of Israel. [29] And everyone who has left houses or brothers or sisters or father or mother or children or fields for my sake will receive a hundred times as much and will inherit eternal life. [30] But many who are first will be last, and many who are last will be first.

We will partake of the same cup, as Jesus.

Is that cup only for Jew's? Of course not. It is for the New Israel of a better Covenant. You and me. By the grace of God, and that only.

Matthew 20:20 (NIV)

A Mother's Request

20:20-28pp—Mk 10:35-45

[20] Then the mother of Zebedee's sons came to Jesus with her sons and, kneeling down, asked a favor of him.

[21] "What is it you want?" he asked.

She said, "Grant that one of these two sons of mine may sit at your right and the other at your left in your kingdom."

[22] "You don't know what you are asking," Jesus said to them. "Can you drink the cup I am going to drink?"

"We can," they answered.

[23] Jesus said to them, "You will indeed drink from my cup, but to sit at my right or left is not for me to grant. These places belong to those for whom they have been prepared by my Father."

God did not fail with Israel! His Israel is different now. It is a New Covenant, that we now are part of. And thankfully, all the promises remain.

This New Covenant is better in every way, and it includes, all those, which have struggled, with God and man Israel!

There is no separate kingdom of Jews, that is different or parallel to a gentiles kingdom! There is a transition time, which we now live in. Where there still is a physical existing nation of Israel, with Jerusalem, and the ancient temple wall. But that is fading,

Hebrews 8:13 By calling this covenant "new," he has made the first one obsolete; and what is obsolete and aging will soon disappear.

Do you say this covenant spoken of in Hebrews 8, is only for Jews? To me, that is blind and ridiculous.

I hope does not offend any, but it is what I feel I must say. The Prince of the power of the air, does not want us to know who we are.

We are still physical and only have a down-payment of the Holy Spirit from our conception (our baptism). When Christ returns, at the sound of a great trumpet, we shall all be changed into spiritual beings. We will see God as He is for we shall be like-Him!

Together we are all the temple, and after this transition period, when our change has fully come, there is neither Jew, nor Greek, male nor female, but only the oneness of spirit Family of God.

> **Matthew Henry's Concise Commentary on the Whole Bible** — Psalm 102:12-22
>
> Ps 102:12-22 We are dying creatures, but God is an everlasting God, the protector of his church; we may be confident that it will not be neglected. When we consider our own vileness, our darkness and deadness, and the manifold defects in our prayers, we have cause to fear that they will not be received in heaven; but we are here assured of the contrary, for we have an Advocate with the Father, and are under grace, not under the law. Redemption is the subject of praise in the Christian church; and that great work is described by the temporal deliverance and restoration of Israel. Look down upon us, Lord Jesus; and bring us into the glorious liberty of thy children, that we may bless and praise thy name.

Figure 43. *Future Restoration of Israel (Matthew Henry's Concise Commentary on the Whole Bible)*

But You, LORD, are enthroned forever; Your fame endures to all generations. You will rise up and have compassion on Zion, for it is time to show favor to her —the appointed time has come. For Your servants take delight in its stones and favor its dust.

Then the nations will fear the name of Yahweh, and all the kings of the earth Your glory, for the LORD will rebuild Zion; He will appear in His glory. He will pay attention to the prayer of the destitute and will not despise their prayer.

This will be written for a later generation, **and a newly created people will praise the LORD***: He looked down from His holy heights—the LORD gazed out from heaven to earth— to hear a prisoner's groaning, to set free those condemned to die, so that they might declare the name of Yahweh in Zion and His praise in Jerusalem, when peoples and kingdoms are assembled to serve the LORD.*

He has broken my strength in midcourse; He has shortened my days. I say: "My God, do not take me in the middle of my life! ***Your years continue through all generations. Long ago You established the earth, and the heavens are the work of Your hands. They will per-***

ish, but You will endure; all of them will wear out like clothing. You will change them like a garment, and they will pass away. But You are the same, and Your years will never end. Your servants' children will dwell securely, and their offspring will be established before You." (Psalms 102:12–27, HCSB)

I believe this is now true of the New Jerusalem as well, for God did not fail. The promises from God will now be fulfilled because Jesus has gotten God's people back on track to the Father—the new covenant people of God, physical and spiritual sons of Abraham, the bride of Christ ISRAEL.

Galatians 6:15 (KJV) For in Christ Jesus neither circumcision availeth any thing, nor uncircumcision, but a new creature.

[16] And as many as walk according to this rule, peace be on them, and mercy, and upon the Israel of God.

[17] From henceforth let no man trouble me: for I bear in my body the marks of the Lord Jesus.

[18] Brethren, the grace of our Lord Jesus Christ be with your spirit. Amen. (Unto the Galatians written from Rome.)

We forget **we are part of Gods spiritual temple,** and some actually hate the notion that we are still Israel, benefactors of God's gracious promises. **Please remember, the Old Testament was replaced but not the promises.** The promises still stand. Unless the covenant with night and day is gone, then and only then would ever Israel be gone. So says the Lord!

Israel—us—has a new and better covenant, with the same inheritance, to be part of the God family—in the glory of space!

"The time is coming," declares the LORD,
**"when I will make a new covenant
with the house of Israel
and with the house of Judah.**
*It will not be like the covenant
I made with their forefathers
when I took them by the hand*

to lead them out of Egypt,
because they broke my covenant,
though I was a husband to them,"
declares the LORD.
"This is the covenant I will make with the house of Israel
after that time," declares the LORD.
"I will put my law in their minds
and write it on their hearts.
I will be their God,
and they will be my people.
No longer will a man teach his neighbor,
or a man his brother, saying, 'Know the LORD,'
because they will all know me,
from the least of them to the greatest,"
declares the LORD.
"For I will forgive their wickedness
and will remember their sins no more."
This is what the LORD says,
he who appoints the sun
to shine by day,
who decrees the moon and stars
to shine by night,
who stirs up the sea
so that its waves roar—
the LORD Almighty is his name:
==**"Only if these decrees vanish from my sight,"**==
==**declares the LORD,**==
==**"will the descendants of Israel ever cease**==
==**to be a nation before me."**==
This is what the LORD says:
"Only if the heavens above can be measured
and the foundations of the earth below be searched out
will I reject all the descendants of Israel
because of all they have done,"
declares the LORD. *(Jeremiah 31:31–37, NIV)*

Praise be to the LORD, the God of Israel,
from everlasting to everlasting.
Let all the people say, "Amen!"
Praise the LORD. (Psalms 106:48, NIV)

And from the New Testament regarding the future of Israel: Israel—us—have a new and better covenant, with the same inheritance.

God did not fail; it will be as He said from the beginning! It was us who failed. Now Christ is responsible for Israel and our foundation, which will never fail!

But God found fault with the people and said:
"The time is coming, declares the Lord,
when **I will make a new covenant**
with the house of Israel
and with the house of Judah.
It will not be like the covenant
I made with their forefathers
when I took them by the hand
to lead them out of Egypt,
because they did not remain faithful to my covenant,
and I turned away from them,
declares the Lord.
[10] This is the covenant I will make with the house of Israel
after that time, declares the Lord.
I will put my laws in their minds
and write them on their hearts.
I will be their God,
and they will be my people.
No longer will a man teach his neighbor,
or a man his brother, saying, 'Know the Lord,'
because they will all know me,
from the least of them to the greatest.
For I will forgive their wickedness

and will remember their sins no more."
(Hebrews 8:8–12, NIV)

I hope that we can now see, and no longer have forgotten, we are Israel, rulers with God! Forgetfulness happened in ancient times. Here is a precedent of our poor memory issues from ancient Judah.

In 2 Kings 22, there is an almost unbelievable story about how the book of the law **had been lost from Judah** and lost in the very temple itself. This was discovered by one of the greatest kings of Judah, King Josiah, and the high priest Hilkiah. If I understand it correctly, for a period of almost three hundred years, Israel had not been keeping the Passover, like directed in the book of the law. Canada is only 150 years old in 2017, and so it is amazing this could be true of forgetful ancient Judah for so long.

> *The Book of the Law Found*
> *Hilkiah the high priest told Shaphan the court secretary,* ==*"I have found the book of the law in the LORD's temple,"*== *and he gave the book to Shaphan, who read it. Then Shaphan the court secretary went to the king and reported, "Your servants have emptied out the money that was found in the temple and have put it into the hand of those doing the work—those who oversee the LORD's temple." Then Shaphan the court secretary told the king, "Hilkiah the priest has given me a book," and Shaphan read it in the presence of the king.* ==*When the king heard the words of the book of the law, he tore his clothes*== *Then he commanded Hilkiah the priest, Ahikam son of Shaphan, Achbor son of Micaiah, Shaphan the court secretary, and the king's servant Asaiah: "Go and inquire of the LORD for me, the people, and all Judah about the instruction in this book that has been found. For great is the LORD's wrath that is kindled against us because* ==**our ancestors**== **have not obeyed the words of this book** *in order to do everything written about us." (2 Kings 22:8–13, HCSB)*

So we see that forgetfulness is not unique to us, like ancient Judah forgot. But as we see from the above example, even if man forgets who he is, **God does not forget**. And when we join up with Christ, when we belong to Christ, we are grafted in as descendants of Abraham, heirs to the eternal promises and eternal purpose of our God.

Please remember with renewed honor Christ, our Passover, sacrificed for us—now!

> *Your glorying is not good. Know ye not that a little leaven leaveneth the whole lump?*
> *Purge out therefore the old leaven, that ye may be a new lump, as ye are unleavened.* For even **Christ our passover is sacrificed for us:**
> *Therefore let us keep the feast, not with old leaven, neither with the leaven of malice and wickedness; but with the unleavened bread of sincerity and truth. (1 Corinthians 5:6–8, KJV)*

Christ, of the New Testament, He is our Passover! As it says in First Corinthians, He is the Passover for Christians, not just people of the Old Testament under the ex-law of Moses.

This is a thing for both testaments, the Passover! Also Pentecost (count 50), the feast of firstfruits and the day of atonement. Passover came from the feasts of the Lord, and so did the day of atonement. If you deny atonement applies for Christians today, then you deny the precedent of the Passover applies to us.[4]

> *For if, when we were enemies, we were reconciled to God by the death of his Son, much more, being reconciled, we shall be saved by his life.*
> *And not only so, but we also joy in God through our Lord Jesus Christ, by whom we have now received the* **atonement**. *(Romans 5:10–11, KJV)*

> **CHRIST OUR ATONEMENT**
>
> On the Day of Atonement described in Leviticus 16, the sins and transgressions of the people are literally placed on the goat of removal and taken away. Jesus Christ, our Atonement, not only provides the entire removal of our sin and sin nature but also imparts back to us His divine nature and righteousness. We determine this Day of Atonement concludes with our full appropriation of the total sanctification provided by the Lamb of God.

Figure 44. At-One With Christ (eclectic Google search)

These are part of God's feasts, but we, of course, have forgotten those and replaced them with evergreen trees and yule logs reflecting the birth of the sun in the winter solstice and bunny rabbits and hot cross buns about fertility and Ishtar the Queen of Heaven. No one denies Leviticus 23 outlines the **feast Days of God for His people, Israel**. It is just that we have been told to say that they are only for the Old Testament.

That is not true. **Yes, the law of Moses is replaced with the law of Christ.** However, just like the promises are forever, so too are the means to obtaining them. **Not through the law of Moses, but newly through Christ our Passover, who always would have to die for Israel to make atonement with the Father.** That stands eternal, covering both covenants. The old is passing over to the new. Only the remnants are left on earth, as we draw nearer to our promised spiritual birth when we fully become part of the family of God and see God as He is in full spiritual glory! This spiritual birth is completed at the return of Christ, at the last trump, the "mystery" previously hidden.

Again, Christ is **our Passover** that brings Christians to be **at one with God**. And we are to become the **firstfruits of the promises**. We are the **New Testament Israel that will enter a gate with the tribes of Israel proudly represented in the temple of the New Jerusalem. And we are the people whom these ancient days are fulfilled upon**. Man, although we promised, could not live up to this, but the "God who would-be man" Christ could and did! It is astounding that Christianity has (for a time) forgotten this or has been blinded to it. It takes a great force to accomplish that. Perhaps we have been deceived and forgotten, but God has not forgotten exactly who we are! As modern spiritual Israel of the new covenant, we still need to clean our temple, just like Josiah above.

Here below, from perhaps the newly discovered book of the Lord are the feast days of God. Even if they were only for the Old Testament Israel, <u>they are from God</u>. And due some respect? Should you completely forget them and have Christmas and Easter, <u>words nowhere found in God's word, the Bible</u>?

Do you not find that strange? I think a very powerful lying voice has done a great job of deceiving the whole world and changing times and seasons.[24]

> *The LORD said to Moses, "Speak to the Israelites and say to them:* **'These are my appointed feasts, the appointed feasts of the LORD**, *which you are to proclaim as sacred assemblies.*
>
> ### The Sabbath
> *"'There are six days when you may work, but the seventh day is a Sabbath of rest, a day of sacred assembly. You are not to do any work; wherever you live, it is a Sabbath to the LORD.*
>
> ### The Passover and Unleavened Bread
> *"'These are the LORD's appointed feasts, the sacred assemblies you are to proclaim at their appointed times: The LORD's Passover begins at twilight on the fourteenth day of the first month. On the fifteenth day of that month the LORD's Feast of Unleavened Bread begins; for seven days you must eat bread made without yeast. On the first day hold a sacred assembly and do no regular work. For seven days present an offering made to the LORD by fire. And on the seventh day hold a sacred assembly and do no regular work.'" [cf Exodus 12:14–20; Numbers 28:16–25; Deuteronomy 16:1–8]*

Firstfruits

The LORD said to Moses, "Speak to the Israelites and say to them: 'When you enter the land I am going to give you and you reap its harvest, bring to the priest a sheaf of the first grain you harvest. He is to wave the sheaf before the LORD so it will be accepted on your behalf; the priest is to wave it on the day after the Sabbath. On the day you wave the sheaf, you must sacrifice as a burnt offering to the LORD a lamb a year old without defect, together with its grain offering of two-tenths of an ephah of fine flour mixed with oil—an offering made to the LORD by fire, a pleasing aroma—and its drink offering of a quarter of a hin of wine. You must not eat any bread, or roasted or new grain, until the very day you bring this offering to your God. This is to be a lasting ordinance for the generations to come, wherever you live.

Feast of Weeks

"'From the day after the Sabbath, the day you brought the sheaf of the wave offering, count off seven full weeks. Count off fifty days up to the day after the seventh Sabbath, and then present an offering of new grain to the LORD. From wherever you live, bring two loaves made of two-tenths of an ephah of fine flour, baked with yeast, as a wave offering of firstfruits to the LORD. Present with this bread seven male lambs, each a year old and without defect, one young bull and two rams. They will be a burnt offering to the LORD, together with their grain offerings and drink offerings—an offering made by fire, an aroma pleasing to the LORD. [Then sacrifice one male goat for a sin offering and two lambs, each a year old, for a fellowship offering. [The priest is to wave the two lambs before the LORD as a wave offering, together with the bread of the firstfruits. They are a sacred offering to the LORD for the priest. On that *same day you are to proclaim a sacred assembly and do no regular work. This is to be a lasting ordinance for the generations to come, wherever you live.*

"'When you reap the harvest of your land, do not reap to the very edges of your field or gather the gleanings of your harvest. Leave them for the poor and the alien. I am the LORD your God.'" [cf Numbers 28:26–31; Deuteronomy 16:9–12]

Feast of Trumpets
The LORD said to Moses, "Say to the Israelites: 'On the first day of the seventh month you are to have a day of rest, a sacred assembly commemorated with trumpet blasts. Do no regular work, but present an offering made to the LORD by fire.'" [cf Number 29:1–6]

Day of Atonement
The LORD said to Moses, "The tenth day of this seventh month is the Day of Atonement. Hold a sacred assembly and deny yourselves, and present an offering made to the LORD by fire. Do no work on that day, because it is the Day of Atonement, when atonement is made for you before the LORD your God. Anyone who does not deny himself on that day must be cut off from his people. I will destroy from among his people anyone who does any work on that day. You shall do no work at all. This is to be a lasting ordinance for the generations to come, wherever you live. It is a sabbath of rest for you, and you must deny yourselves. From the evening of the ninth day of the month until the following evening you are to observe your sabbath." [Leviticus 16:2–34; Numbers 29:7–11]

Feast of Tabernacles
The LORD said to Moses, "Say to the Israelites: 'On the fifteenth day of the seventh month the LORD's Feast of

Tabernacles begins, and it lasts for seven days. The first day is a sacred assembly; do no regular work. For seven days present offerings made to the LORD by fire, and on the eighth day hold a sacred assembly and present an offering made to the LORD by fire. It is the closing assembly; do no regular work.

("'These are the LORD's appointed feasts, which you are to proclaim as sacred assemblies for bringing offerings made to the LORD by fire—the burnt offerings and grain offerings, sacrifices and drink offerings required for each day. These offerings are in addition to those for the LORD's Sabbaths and in addition to your gifts and whatever you have vowed and all the freewill offerings you give to the LORD.)

"'So beginning with the fifteenth day of the seventh month, after you have gathered the crops of the land, celebrate the festival to the LORD for seven days; the first day is a day of rest, and the eighth day also is a day of rest. On the first day you are to take choice fruit from the trees, and palm fronds, leafy branches and poplars, and rejoice before the LORD your God for seven days. Celebrate this as a festival to the LORD for seven days each year. This is to be a lasting ordinance for the generations to come; celebrate it in the seventh month. Live in booths for seven days: All native-born Israelites are to live in booths so your descendants will know that I had the Israelites live in booths when I brought them out of Egypt. I am the LORD your God.'" [cf Numbers 29:12–39; Deuteronomy 16:13–17]

So Moses announced to the Israelites the appointed feasts of the LORD. *(Leviticus 23, NIV)*

The Lord does not forget, and so we are the not-forgotten Israel of Abraham's seed.

SPACE THE TRUE FRONTIER!

Shout for joy, O heavens;
rejoice, O earth;
burst into song, O mountains!
For the LORD comforts his people
and will have compassion on his afflicted ones.
But Zion said, "The LORD has forsaken me,
the Lord has forgotten me."
"Can a mother forget the baby at her breast
and have no compassion on the child she has borne?
Though she may forget,
I will not forget you! *(Isaiah 49:13–15, NIV)*

If you belong to Christ, then you are Abraham's seed, and heirs according to the promise.
(Galatians 3:29, NIV)

Wherefore, remember, that ye were once the nations in the flesh, who are called Uncircumcision by that called Circumcision in the flesh made by hands, that ye were at that time apart from Christ, having been alienated ***from the commonwealth of Israel, and strangers to the covenants of the promise****, having no hope, and without God, in the world; and now, in Christ Jesus, ye being once afar off became nigh in the blood of the Christ. (Ephesians 2:11–13, NIV)*

Note regarding "the commonwealth of Israel"

After Jacob returned from Paddan Aram, God appeared to him again and blessed him. God said to him, "Your name is Jacob, but you will no longer be called Jacob; ***your name will be Israel.****" So he named him Israel. And God said to him, "I am God Almighty; be fruitful and increase in number.* ***A nation*** *and* ***a community of nations*** *will come from you, and kings will come from your body. (Genesis 35:9–11, NIV)*

> *Therefore, the promise comes by faith, so that it may be by grace and may* **be guaranteed to all Abraham's offspring**—*not only to those who are of the law but also to those who are of the faith of Abraham.* **He is the father of us all**. *As it is written: "I have made you a father of many nations." He is our father in the sight of God, in whom he believed—the God who gives life to the dead and calls things that are not as though they were. (Romans 4:16–17, NIV)*

In the New Testament, Israel is compared to an olive tree and is shown to include Gentiles (wild olive branches), thereby showing that all are Israel and descendants of the promises that will never fail.

> *After all, if you were cut out of an olive tree that is wild by nature, and contrary to nature were grafted into a cultivated olive tree,* **how much more readily will these, the natural branches, be grafted into their own olive tree!**

> **All Israel Will Be Saved**
> *I do not want you to be ignorant of this mystery, brothers, so that you may not be conceited: Israel has experienced a hardening in part until the full number of the Gentiles has come in.* **And so all Israel will be saved, as it is written:**
> *"The deliverer will come from Zion;*
> *he will turn godlessness away from Jacob.*
> *And this is my covenant with them*
> **when I take away their sins."** *(Romans 11:24–27, NIV)*

Speaking of the **future of us, Israel**:

> *And* **Jesus said** *unto them, Verily I say unto you, that ye which have followed me, in the regeneration* **when the Son of man shall sit in the throne of his glory,**

> *ye also shall sit upon twelve thrones, judging the twelve tribes of Israel.* (Matthew 19:28, KJV)

The future temple of God has twelve gates that we all will walk through. They have pillars with the names of the twelve apostles and on the gates the names of the twelve tribes of Israel!

> *It had a great, high wall with twelve gates, and with twelve angels at the gates.* **On the gates were written the names of the twelve tribes of Israel.** *[13] There were three gates on the east, three on the north, three on the south and three on the west. [14] The wall of the city had twelve foundations,* **and on them were the names of the twelve apostles of the Lamb.** *(Revelation 21:12–14, NIV)*

Don't forget we are Israel, the firstfruits of the harvest of God from the field He has sown. And the promises are for us. We are part of the temple in the New Jerusalem. We will rule with Christ over the twelve tribes of Israel.

The means to get us there has never changed. Christ our Passover has brought us to be at one with the Father through the wedding supper of the Lamb.

Proof of part in that temple:

We are a building, part of the temple of God, if we have Christ and the Father in us.

> *I did not see a temple in the city, because* **the Lord God Almighty and the Lamb are its temple.** *The city does not need the sun or the moon to shine on it, for* **the glory of God gives it light, and the Lamb is its lamp.** *The nations will walk by its light, and the kings of the earth will bring their splendor into it. On no day will its gates ever be shut, for there will be no night there. The glory and honor of the nations will be brought into it. Nothing impure will ever enter it, nor will anyone*

who does what is shameful or deceitful, but only those **whose names are written in the Lamb's book of life.** *(Revelation 21:22–27, NIV)*

What agreement is there between the temple of God and idols? **For we are the temple of the living God. As God has said: "I will live with them and walk among them, and I will be their God, and they will be my people."**
*"Therefore come out from them
and be separate, says the Lord.
Touch no unclean thing,
and I will receive you." (2 Corinthians 6:16–17, NIV)*

Consequently, you are no longer foreigners and aliens, but fellow citizens with God's people **and members of God's household, built on the foundation of the apostles and prophets**, *with* **Christ Jesus himself as the chief cornerstone. In him the whole building is joined together and rises to become a holy temple in the Lord. And in him you too are being built together to** become a dwelling in which God lives by his Spirit. *(Ephesians 2:19–22, NIV)*

Wow! The temple is the spiritually composed family of God—Father, Son, and bride! Don't forget it because God has not forgotten. And please notice, there is no third spirit-person in the true temple of God.

I say again, beware of this third person (spirit) antichrist that appears as an angel of light at the temple, inserting himself as God-seeking your worship. Just as he tried to deceive Christ, he tries to deceive us into believing he is God, (he is not). The Trinity, is NOT.

Concerning the coming of our Lord Jesus Christ and our being gathered to him, we ask you, brothers, [2] not to become easily unsettled or alarmed by some prophecy, report or letter supposed to have come from us, saying that the day of the Lord has already come. **Don't let anyone deceive you in any way, for [that day will not come] until the rebellion occurs and the man of lawlessness is revealed,** *the man doomed to destruction.* **He will oppose and will exalt himself over everything that is called God or is worshiped, so that he sets himself up in God's temple, proclaiming himself to be God.**
Don't you remember that when I was with you I used to tell you these things? And now you know what is holding him back, so that he may be revealed at the proper time. For the secret power of lawlessness is already at work; but the one who now holds it back will continue to do so till he is taken out of the way. And then the lawless one will be revealed, **whom the Lord Jesus will overthrow with the breath of his mouth** *and destroy by the splendor of his coming. The coming of the lawless one will be in accordance with the work of Satan displayed in all kinds of counterfeit miracles, signs and wonders, and in every sort of evil that deceives those who are perishing. They perish because they refused to love the truth and so be saved. For this reason* **God sends them a powerful delusion so that they will believe the lie.** *(2 Thessalonians 2:1–11, NIV)*

God is not a Trinity, or three-person Godhead. Ultimately, God will be a spiritual life essence consisting of Father, Son, and bride (atonement). I hope you read it and rejoice. Why would you remain blind? I hope God has given you this gift of sight. Open your eyes and ears, and understand! Throw away the idols, and throw away the hate, of the lying voice. Get behind us, Satan! You are not God, you are not part of the temple!

I hope you can **remember** who you are Israelite and thank God. What an awesome God, and what a glorious future He is sharing! The spacious firmament, our future home, is closer now.

Yes, I am SHOUTING in this book because I feel like a fog light on a stormy foggy night that has lasted millennia. I must remember, only God can lift the fog, and I pray He does.

CHAPTER 5

KNOCK, KNOCK!

If I were to say "Knock, knock" to any child, I am sure I would hear them reply, "Who's there?" Please keep that thought in mind as you read on. Think, "Who's there?"

This chapter is intended to show that because the Holy Spirit is not a person, it can be left out of many discussions (or treated differently) where the Godhead, or the temple of God, is involved. In fact, many of the books of the New Testament have introductions that exclude the Spirit. It would be an embarrassment to leave out the Holy Spirit if "it" was an equal third person "he."

I am not saying there is no Holy Spirit; I am just saying it is not one of the personalities of the God Head. Spirit is what they are made of. That's why it gets treated differently in these examples. Let me show you what I am talking about.

Example 1, from Romans:

> *Paul, a servant of Jesus Christ, a called apostle, having been separated to the good news of* **God**—*which He announced before through His prophets in holy writings—concerning His Son, (who is come of the seed of David according to the flesh, who is marked out* **Son of God** *in* **power, according to the Spirit** *of sanctification, by the rising again from the dead,) Jesus Christ our Lord; through whom we did receive grace and apostleship, for obedience of faith among all the nations, in behalf of his name; among whom are also ye, the called of Jesus Christ; to all who are in Rome, beloved of God, called saints;* **Grace to you, and peace, from God our Father, and from the Lord Jesus Christ!** *(Romans 1:1–7, YLT)*

In the opening sentences of the book of Romans, we can say "Knock, knock" and answer the Father of Jesus (God), the Son of God (Jesus), and power-in-spirit form are there. But especially notice the last verse quoted and that only God the Father and the Lord Jesus Christ are mentioned as personalities. Amazing true?

Example 2, from First Corinthians:

> *Paul, a called apostle of Jesus Christ, through the will of God, and Sosthenes the brother, to the assembly of God that is in Corinth, to those sanctified in Christ Jesus, called saints, with all those calling upon the name of our Lord Jesus Christ in every place--both theirs and ours: Grace to you and peace from God our Father and the Lord Jesus Christ! (1 Corinthians 1:1–3, YLT)*

In these first few sentences of First Corinthians, we find the will of God—God the Father and the Lord Jesus Christ. Same thing, the last statement only involves the persons God the Father and the Lord Jesus. **The will of God is how the Holy Spirit is represented because the Spirit is that blessed will.**

Example 3, from Ephesians 1:

> *Paul, an apostle of Jesus Christ through the will of God, to the saints who are in Ephesus, and to the faithful in Christ Jesus: Grace to you, and peace from God our Father, and the Lord Jesus Christ!*
> *Blessed is the God and Father of our Lord Jesus Christ, who did bless us in every spiritual blessing in the heavenly places in Christ. (Ephesians 1:1–3, YLT)*

Amazing. Again the persons of God the Father, the Son Jesus, and the will of God.

Example 4, from first Peter:

> *Peter, an apostle of Jesus Christ, to the choice sojourners of the dispersion of Pontus, Galatia, Cappadocia, Asia,*

> *and Bithynia, according to a foreknowledge of God the Father, in sanctification of the Spirit, to obedience and sprinkling of the blood of Jesus Christ: Grace to you and peace be multiplied!*
> *Blessed is the God and Father of our **Lord Jesus Christ**, who, according to the abundance of His kindness did beget us again to a living hope, through the rising again of Jesus Christ out of the dead, to **an inheritance incorruptible, and undefiled, and unfading, reserved in the heavens for you**, who, in the power of God are being guarded, through faith, unto salvation, ready to be revealed in the last time. (1 Peter 1:1–5, YLT)*

In the example above, the sanctifying spirit is mentioned, but it is not a person. It is shown to be the sanctifying power of God that protects us in our faith. I ask you, if this will of God doesn't represent the nonperson component of God, **then where is the third person?**

Example 5, from Galatians:

> *Paul, an apostle—not from men, nor through man, but through Jesus Christ, and God the Father, who did raise him out of the dead—and all the brethren with me, to the assemblies of Galatia: Grace to you, and peace from God the Father, and our Lord Jesus Christ, who did give himself for our sins, that he might deliver us out of the present evil age, according to the will of God even our Father, to whom is the glory to the ages of the ages. Amen. (Galatians 1:1–5, YLT)*

Through Jesus, the Father, and according to the will of the Father (Spirit).

Example 6, from 1 John:

> *That which was from the beginning, that which we have heard, that which we have seen with our eyes, that which we did behold, and our hands did handle, con-*

> *cerning the Word of the Life—and the Life was manifested, and we have seen, and do testify, and declare to you the Life, the age-during, which was with the Father, and was manifested to us—that which we have seen and heard declare we to you, that ye also may have fellowship with us, and our **fellowship** is with the Father, and with His Son Jesus Christ; and these things we write to you, that your joy may be full. (1 John 1:1–4, YLT)*

This, to me, is the best example of where the spirit can be excluded from the discussion because it is there by default. It is what God is made of (John 4:24, 2 Corinthians 3:17, 18). As mentioned before, it could just have easily been included because it is what they are made of. *"The grace of the Lord Jesus Christ, and the love of God, and the **fellowship** of the Holy Spirit, is with you all! Amen" (2 Corinthians 13:14, YLT).*

Trinitarians focus on 2 Corinthians 13:14 and seem to ignore 1 John 1:1–4. Non-Trinitarians want to ignore 2 Corinthians 13:14 (that used to be me). We seem to just want to see things that back up our image (but that is idolatry-worshipping man-made images).

God is what God is, and we should humbly listen to what He says! My heart tells me—and I think it is from God—that the fact the spirit can be left out of or included in this discussion of our **fellowship** is the major point.

It is only possible because there are two personalities and the seal of the spirit involved! If you choose to focus on personalities, then you don't mention spirit; if you want to focus on the fact that these persons are composed of spirit, then you include it.

It would not be possible to talk about our **fellowship** and leave the spirit out, if it were an equal third person. When you understand that the Father and the Son are made of spirit, you know that essence is there by default, and so it is okay not to mention the obvious (the spirit essence is always there).

In 2 Corinthians 13:14, Paul decided to include the obvious and mention spirit presence, but it cannot be a person, or it

would also have to be mentioned in First John, where our joy is complete in the fellowship of Father and Son.

I now realize that First John, along with Second Corinthians, and their synergy, makes a much clearer understanding of my point. These scriptures are not a contradiction to each other, they clarify.

Example 7, from Jude:

> *Judas, of Jesus Christ a servant, and brother of James, to those sanctified in God the Father, and in Jesus Christ kept--called, [2] kindness to you, and peace, and love, be multiplied! (Jude 1, YLT)*

Once again, leaving out the spirit is okay because it is not a person; it is, however, there by default as mentioned earlier. I have shown from these examples that, there is a Father, a Son (Jesus), and there is the Holy Spirit, but only two of these are personalities. The Spirit is the powerful composition, or will, of God that is spoken of separately but is sometimes left out when personalities are involved.

The Holy Spirit is there; it is the works of God, his will in action, but Christ and the Father are the personalities that make up that Spirit. They are the Spirit (2 Corinthians 3:17).

Final example that shows this distinction from Second Thessalonians:

> *Stand Firm*
> *But we ought always to thank God for you, brothers loved by the Lord, because from the beginning God chose you to be **saved through the sanctifying work of the Spirit** and through belief in the truth. He called you to this through our gospel, that you might share in the glory of our Lord Jesus Christ. So then, brothers, stand firm and hold to the teachings we passed on to you, whether by word of mouth or by letter. (2 Thessalonians 2:13, NIV)*
>
> ***May our Lord Jesus Christ himself and God our Father**, who loved us and by his grace gave us eternal*

encouragement and good hope, encourage your hearts and strengthen you in every good deed and word. (2 Thessalonians 2:13–17, NIV)

Request for Prayer
Finally, brothers, pray for us that the message of the Lord may spread rapidly and be honored, just as it was with you. And pray that we may be delivered from wicked and evil men, for not everyone has faith. But the Lord is faithful, and he will strengthen and protect you from the evil one. We have confidence in the Lord that you are doing and will continue to do the things we command. May the Lord direct your **hearts into God's love and Christ's perseverance**. *(2 Thessalonians 3:1–5)*

I think these examples show that there is the Father, the Son, and the Holy Ghost, but not in three persons (Trinity). The Spirit is different; it is the powerful working will of God in action. It is the living water that flows out of Christ the Lamb from the throne of God. It is passed on by the laying on of hands from God to God's people and through God's people (John 7:37–39, Revelation 22:1–2)

Next, I would like to move on to what I call the book of atonement. Where is that you might ask? Well, it is in the New Testament of the Bible, and you have probably read it many times, but if you're like me, it takes some time to realize the full light of its meaning. We are to become at-one with God, through the work of Christ our atonement! Wow!

But first **an introduction to the book of atonement**:

Christ's Sacrifice Once for All
The law is only a shadow of the good things that are coming—not the realities themselves. For this reason it can never, by the same sacrifices repeated endlessly year after year, make perfect those who draw near to worship. If it could, would they not have stopped being offered? For the worshipers would have been cleansed once for

> *all, and would no longer have felt guilty for their sins. But those sacrifices are an annual reminder of sins, [because it is impossible for the blood of bulls and goats to take away sins.*
> *Therefore, when Christ came into the world, he said: "Sacrifice and offering you did not desire,*
> *but a body you prepared for me;*
> *with burnt offerings and sin offerings*
> *you were not pleased.*
> *Then I said, 'Here I am—it is written about me in the scroll—*
> *I have come to do your will, O God.'"*
> *First he said, "Sacrifices and offerings, burnt offerings and sin offerings you did not desire, nor were you pleased with them" (although the law required them to be made). Then he said, "Here I am, I have come to do your will." He sets aside the first to establish the second. And by that will, we have been made holy through the sacrifice of the body of Jesus Christ once for all. (Hebrews 10:1–10, NIV)*
>
> *For if, when we were enemies, we were reconciled to God by the death of his Son, much more, being reconciled, we shall be saved by his life.*
> *And not only so, but we also joy in God through our Lord Jesus Christ, by whom we have now received the atonement. (Romans 5:10–11, KJV)*

Christ is the ultimate (one-time) fulfillment sacrifice that was previously only foreshadowed by the typical sacrifices of old, pictured in the day of Atonement. There was never any doubt that this true ultimate sacrifice would be needed. It was known since before creation. He was slain as our Passover to atone us with the Father since the foundation of the world. Pictured in the feast days of God, not man's feast days.

Christ is the Passover Lamb, whose blood makes atonement for us with God. God's feast days pictured that. (Note: Christ probably died at the exact time the Jews were slaughtering the Passover lambs by the temple that year. What an awesome fulfillment).[14]

> *Your glorying is not good. Know ye not that a little leaven leaveneth the whole lump?*
> *Purge out therefore the old leaven, that ye may be a new lump, as ye are unleavened. For even* **Christ our passover** *is sacrificed for us:*
> *Therefore let us keep the feast, not with old leaven, neither with the leaven of malice and wickedness; but with the unleavened bread of sincerity and truth. (1 Corinthians 5:6–8, KJV)*

Below I have included a good summary of God's feast of atonement and how in the New Testament fulfilment Christ became our Passover atonement Lamb. I did a Google search, and eclectically found this statement which echo's what I am saying.[25]

Figure 45. At One With Christ Repeated (eclectic Google search)

CHRIST OUR ATONEMENT

On the Day of Atonement described in Leviticus 16, the sins and transgressions of the people are literally placed on the goat of removal and taken away. Jesus Christ, our Atonement, not only provides the entire removal of our sin and sin nature but also imparts back to us His divine nature and righteousness. We determine this Day of Atonement concludes with our full appropriation of the total sanctification provided by the Lamb of God.

Okay, it's not a book of atonement, it is just a chapter, but it is totally revealing on this matter of "Knock, Knock." After Jesus had accomplished all he set out to do, he prays to the Father. He knows his work is done, and he will be gone soon, so he imparts a blessing on his disciples and those who follow them (including us). It is the chapter of atonement, John 17, my favorite chapter in the Bible.

"Knock! Knock!"

"Who's there?"

SPACE THE TRUE FRONTIER!

These things spake Jesus, and lifted up his eyes to the heaven, and said—"Father, the hour hath come, glorify Thy Son, that Thy Son also may glorify Thee, according as Thou didst give to him authority over all flesh, that--all that Thou hast given to him—he may give to them life age-during; and this is the life age-during, that they may know Thee, the only true God, and him whom Thou didst send—Jesus Christ; I did glorify Thee on the earth, the work I did finish that Thou hast given me, that I may do it. 'And now, glorify me, Thou Father, with Thyself, with the glory that I had before the world was, with Thee;

I did manifest Thy name to the men whom Thou hast given to me out of the world; Thine they were, and to me Thou hast given them, and Thy word they have kept; now they have known that all things, as many as Thou hast given to me, are from Thee, because the sayings that Thou hast given to me, I have given to them, and they themselves received, and have known truly, that from Thee I came forth, and they did believe that Thou didst send me. 'I ask in regard to them; not in regard to the world do I ask, but in regard to those whom Thou hast given to me, because Thine they are, and all mine are Thine, and Thine are mine, and I have been glorified in them;

and no more am I in the world, and these are in the world, and I come unto Thee. ==Holy Father, keep them in Thy name, whom Thou hast given to me, that they may be one as we;== *when I was with them in the world, I was keeping them in Thy name; those whom Thou hast given to me I did guard, and none of them was destroyed, except the son of the destruction, that the Writing may be fulfilled. And now unto Thee I come, and these things I speak in the world, that they may have my joy fulfilled in themselves; I have given to them Thy word, and the world did hate them, because they are not of the world, as I am not of the world; I do not ask that Thou mayest take them out of the world, but that Thou mayest keep them out of the evil. 'Of the world they are not, as I of the world am not;*

sanctify them in Thy truth, Thy word is truth; as Thou didst send me to the world, I also did send them to the world; [19] and for them do I sanctify myself, that they also themselves may be sanctified in truth.

'And not in regard to these alone do I ask, but also in regard to those who shall be believing, through their word, in me; **that they all may be one, as Thou Father art in me, and I in Thee; that they also in us may be one***, that the world may believe that Thou didst send me. 'And I, the glory that thou hast given to me, have given to them,* **that they may be one as we are one; I in them, and Thou in me, that they may be perfected into one***, and that the world may know that Thou didst send me, and didst love them as Thou didst love me. Father, those whom Thou hast given to me, I will that where I am they also may be with me, that they may behold my glory that Thou didst give to me, because Thou didst love me before the foundation of the world. ['Righteous Father, also the world did not know Thee, and I knew Thee, and these have known that Thou didst send me, and I made known to them Thy name, and will make known, that the love with which Thou lovedst me in them may be, and I in them.'" (John 17, YLT)*

Once again in a modern translation:

Jesus Prays for Himself
After Jesus said this, he looked toward heaven and prayed:

"Father, the time has come. Glorify your Son, that your Son may glorify you. For you granted him authority over all people that he might give eternal life to all those you have given him. Now this is eternal life: that they may know you, the only true God, and Jesus Christ, whom you have sent. I have brought you glory on earth by completing the work you gave me to do. And now, Father, glorify me in your presence with the glory I had with you before the world began.

Jesus Prays for His Disciples

"I have revealed you to those whom you gave me out of the world. They were yours; you gave them to me and they have obeyed your word. Now they know that everything you have given me comes from you. For I gave them the words you gave me and they accepted them. They knew with certainty that I came from you, and they believed that you sent me. I pray for them. I am not praying for the world, but for those you have given me, for they are yours. All I have is yours, and all you have is mine. And glory has come to me through them. I will remain in the world no longer, but they are still in the world, and I am coming to you. Holy Father, protect them by the power of your name—the name you gave me—so that they may be one as we are one. While I was with them, I protected them and kept them safe by that name you gave me. None has been lost except the one doomed to destruction so that Scripture would be fulfilled.
"I am coming to you now, but I say these things while I am still in the world, so that they may have the full measure of my joy within them. I have given them your word and the world has hated them, for they are not of the world any more than I am of the world. My prayer is not that you take them out of the world but that you protect them from the evil

one. They are not of the world, even as I am not of it. Sanctify them by the truth; your word is truth. As you sent me into the world, I have sent them into the world. For them I sanctify myself, that they too may be truly sanctified.

Jesus Prays <u>for All Believers</u>

"My prayer is not for them alone. I pray also for those who will believe in me through their message, **<u>that all of them may be one, Father, just as you are in me and I am in you. May they also be in us</u>** so that the world may believe that you have sent me. I have given them the glory that you gave me, that they may be one as we are one: I in them and you in me. May they be brought to complete unity to let the world know that you sent me and have loved them even as you have loved me.
"Father, I want those you have given me to be with me where I am, and to see my glory, the glory you have given me because you loved me before the creation of the world.
"Righteous Father, though the world does not know you, I know you, and they know that you have sent me. I have made you known to them, and will continue to make you known in order that the love you have for me may be in them and that I myself may be in them." (John 17:1, NIV)

How could a separate (third person) be left out of this chapter? This is talking about us becoming part of a oneness (atonement) with God—Father, Son, and bride.

If the Holy Spirit was a person in the Godhead, it simply could not be left out! Amazing when you can see it. What a spell we have been under to believe that. There is no third person in this oneness!! Reread John 17, the imposter person is not there.

SPACE THE TRUE FRONTIER!

Hopefully, we reject the imposter as Christ did. There is a third spirit in heaven, making himself out to be God but is not.

> *Don't let anyone deceive you in any way, for [that day will not come] until the rebellion occurs and the man of lawlessness is revealed, the man doomed to destruction.* ==He will oppose and will exalt himself over everything that is called God or is worshiped, so== **that he sets himself up in God's temple, proclaiming himself to be God.** *(2 Thessalonians 2:3–4, NIV)*

That is the same spirit that tried to get Christ to worship him, as discussed earlier. The antichrist, Satan, is not part of the family of God. He is not part of the Godhead. There are no three persons in the Godhead. I hope that is clear by now.

And now a wonderful example from the book of Hebrews showing what the Godhead is made of. Knock, knock! The powerful word of God (His Spirit) and the personalities—the Son and the Father. Please note the Son is worshipped, which is something reserved for God alone**, thereby proving Jesus is God.**

> *The Son Superior to Angels*
> *In the past God spoke to our forefathers through the prophets at many times and in various ways, but in these last days he has spoken to us by* **his Son,** *whom he appointed heir of all things, and through whom he made the universe. [The Son is the radiance of God's glory and the exact representation of his being, sustaining all things by* **his powerful word**. *After he had provided purification for sins, he sat down at* **the right hand of the Majesty in heaven**. *So he became as much superior to the angels as the name he has inherited is superior to theirs.*
> *For to which of the angels did God ever say,*
> **"You are my Son;**
> **today I have become your Father"?**
> *Or again,*

*"I will be his Father,
and he will be my Son"?
And again, when God brings his firstborn into the world, he says,*
"Let all God's angels worship him."
*In speaking of the angels he says,
"He makes his angels winds,
his servants flames of fire."
But about the Son he says,
"Your throne, O God, will last for ever and ever,
and righteousness will be the scepter of your kingdom.
You have loved righteousness and hated wickedness;
therefore God, your God, has set you above your companions
by anointing you with the oil of joy."
He also says,*
**"In the beginning, O Lord, you laid the foundations of the earth,
and the heavens are the work of your hands.**
*They will perish, but you remain;
they will all wear out like a garment.
You will roll them up like a robe;
like a garment they will be changed.
But you remain the same,
and your years will never end."*
**To which of the angels did God ever say,
"Sit at my right hand
until I make your enemies
a footstool for your feet"?
Are not all angels ministering spirits sent to serve those who will inherit salvation?** *(Hebrews 1, NIV)*

A new point now: I have heard a few people pray to the "Holy Spirit," considering it as the third person of the Godhead. But what does Jesus reveal? **Whom do you pray to and by what authority? We pray to the Father.**

> *After this manner therefore pray ye:* **Our Father which art in heaven, Hallowed be thy name.**
> *Thy kingdom come. Thy will be done in earth, as it is in heaven.*
> *Give us this day our daily bread.*
> *And forgive us our debts, as we forgive our debtors.*
> *And lead us not into temptation, but deliver us from evil: For thine is the kingdom, and the power, and the glory, for ever. Amen. (Matthew 6:9–10, KJV)*

And we ask it by the authority of Jesus

> *Believest thou not that I am in the Father, and the Father in me? the words that I speak unto you I speak not of myself: but the Father that dwelleth in me, he doeth the works.*
> **Believe me that I am in the Father, and the Father in me: or else believe me for the very works' sake.**
> *Verily, verily, I say unto you, He that believeth on me, the works that I do shall he do also; and greater works than these shall he do; because I go unto my Father.*
> **And whatsoever ye shall ask in my name, that will I do, that the Father may be glorified in the Son. If ye shall ask any thing in my name, I will do it.**
> *If ye love me, keep my commandments.*
> *John 14:10–15 (KJV)*

The direction from Jesus is to pray to our Father in the name of Jesus. In spiritual form, both Jesus and the Father live in us and bring us comfort for they are the Comforter. Knock, knock!

Romans 8:9-11, New World Translation:

> 9 However, you are in harmony, not with the flesh, but with the spirit, ⁺ if God's spirit truly dwells in you. But if anyone does not have Christ's spirit, this person does not belong to him. 10 But if Christ is in union with you, ⁺ the body is dead because of sin, but the spirit is life because of righteousness. 11 If, now, the spirit of him who raised up Jesus from the dead dwells in you, the one who raised up Christ Jesus from the dead ⁺ will also make your mortal bodies alive ⁺ through his spirit that resides in you.

Figure 46. Christ and the Father, are the combined-Spirit Comforter, Source JW Bible

Our ==Comfort (Counsel)== is that Christ and the Father dwell in us in spirit form. We pray to God the Father by the authority of Jesus, and the powerful wilful spirit of God makes it happen.

It is the will of God that we were foreordained to becoming sons of God, as the bride of Christ. Father Son and bride—a family composed of spirit (no longer any flesh once we are changed to be fully "like" them—spirit).

> *Paul, an apostle of Jesus Christ through the will of God, to the saints who are in Ephesus, and to the faithful in Christ Jesus: Grace to you, and* **peace from God our Father, and the Lord Jesus Christ!**
> *Blessed is the God and Father of our Lord Jesus Christ, who did bless us in every spiritual blessing in the heavenly places in Christ, according as He did choose us in him before the foundation of the world, for our being holy and unblemished before Him, in love, having foreordained us to the adoption of sons through Jesus Christ to Himself, according to the good pleasure of His will, to the praise of the glory of His grace, in which He did make us accepted in the beloved, in whom we have the redemption through his blood, the remission of the trespasses, according to the riches of His grace, in which He*

> *did abound toward us in all wisdom and prudence,* **having made known to us the secret of His will,** *according to His good pleasure, that He purposed in Himself, in regard to the dispensation of the fulness of the times, to bring into one the whole in the Christ, both the things in the heavens, and the things upon the earth—in him;* ***in whom also we did obtain an inheritance, being foreordained*** **according to the purpose of Him who the all things is working according to the counsel of His will,** *for our being to the praise of His glory, even those who did first hope in the Christ, in whom ye also, having heard the word of the truth--the good news of your salvation--in whom also having believed,* ***ye were sealed with the Holy Spirit of the*** promise, *which is an earnest of our inheritance, to the redemption of the acquired possession, to the praise of His glory.* (Ephesians 1:1–14)

What is promised, and we receive as a down payment of our inheritance (by the laying on of hands at baptism), is fulfilled as a mystery in the twinkling of an eye at the last trump.

> *Now it is God who makes both us and you stand firm in Christ. He* **anointed us, set his seal of ownership on us, and put his Spirit in our hearts** *as a deposit,* **guaranteeing what is to come**. *(2 Corinthians 1:21–22, NIV)*

> *I declare to you, brothers, that flesh and blood cannot inherit the kingdom of God, nor does the perishable inherit the imperishable. Listen,* **I tell you a mystery: We will not all sleep, but we will all be changed**—*in a flash,* **in the twinkling of an eye, at the last trumpet**. *For the trumpet will sound, the dead will be raised imperishable, and we will be changed. For the perishable must clothe itself with the imperishable, and the mortal with immortality. When the perishable*

has been clothed with the imperishable, and the mortal with immortality, then the saying that is written will come true: "Death has been swallowed up in victory." (1 Corinthians 15:50–54, NIV)

So will it be with the resurrection of the dead. The body that is sown is perishable, it is raised imperishable; it is sown in dishonor, it is raised in glory; it is sown in weakness, it is raised in power; it is sown a natural body, it is raised a spiritual body.
If there is a natural body, there is also a spiritual body. So it is written: "The first man Adam became a living being"; the last Adam, a life-giving spirit. The spiritual did not come first, but the natural, and after that the spiritual. The first man was of the dust of the earth, the second man from heaven. (1 Corinthians 15:42–47, NIV)

Jesus Prays for All Believers
"My prayer is not for them alone. I pray also for those who will believe in me through their message, that all of them may be one, Father, just as you are in me and I am in you. **May they also be in us so that the world may believe that you have sent me. I have given them the glory that you gave me, that they may be one as we are one:** *I in them and you in me. May they be brought to complete unity to let the world know that you sent me and have loved them even as you have loved me. "Father, I want those you have given me to be with me where I am, and to see my glory, the glory you have given me because you loved me before the creation of the world. "Righteous Father, though the world does not know you, I know you, and they know that you have sent me. (John 17:20–25, NIV)*

Yes, we know who is our true Father, and we know we must remain in—knock, knock—the Father and the son! We are anointed (sealed and protected by the living water spirit) by the essence of the Holy Father.

> *But you have an anointing from the Holy One, and all of you know the truth. I do not write to you because you do not know the truth, but because you do know it and because no lie comes from the truth. Who is the liar? It is the man who denies that Jesus is the Christ. Such a man is the antichrist—*==**he denies the Father and the Son**==. *No one who denies the Son has the Father; whoever acknowledges the Son has the Father also.*
> *See that what you have heard from the beginning remains in you. If it does,* ==**you also will remain in the Son and in the Father.**== *And this is what he promised us—even eternal life. (1 John 2:20, NIV)*

The antichrist denies the Father and the Son; there are the two personalities he denies, not three. We are baptized into the name of the Father into the body of Christ and sealed by Holy Spirit living waters—at one with the family of God.

I know that I have repeated things in this chapter, but it takes time for things to sink in (at least for me). It is hard to overcome millennia of untruth from the great deceiver.

That same third spirit person that inserted himself as god (but is not) has been present in the churches to this day, like Asherah poles in the temple of God in the Old Testament times.

> *"Among the prophets of Samaria*
> *I saw this repulsive thing:*
> *They prophesied by Baal*
> *and led my people Israel astray." (Jeremiah 23:13, NIV)*

==**As King Josiah and the priest Hilkiah (2 Kings 23:1–24) had renewed insight to the purity of God's temple, they tossed out the images of Baal and tore down the Asherah poles. For me this will eventually include the paganism of Christmas and Easter**==

==(Ishtar), which should have no place in my life, according to the Bible, and similarly should not be in God's temple today. So we see that paganism has always crept into the place of true worship, but I believe, like in Josiah's time, paganism will be removed in the future and shown to be the deception it is. God does not use paganism to bring people to him. What does the winter wolstice (rebirth of the sun) evergreen trees, yuletide logs, Easter eggs and bunnies have to do with Christ? Paganism tears people away from God, and He tells us He hates it—always has, always will. In Jeremiah 10:1–12, God says He hates it. We tell Him, "Chill. We use it to help us believe in you, and besides, its pretty and cute." Whose voice are we listening to?==

He tried to gain Christ's worship (acting like his father) and he demanded ours. He said to Christ, "If you just bow down and worship me, you can have those kingdoms, and it will be so much easier you won't have to die."

Satan says to us, "It will be much easier if you just bow down and worship me the third spirit person at the temple of God. Otherwise, I warn you, that you won't have a place to worship in the world if you don't go along with this. Every church has me as part of the Godhead and thereby at least unknowingly worships me. I am the third person of that temple and added to the word of God, what I call a Trinity. In fact, I deny the Father and the Son, so I alone am left to be worshipped as god of the three.

"You can show how I am a rightful part of the temple if you keep what I have changed, the times and seasons. Christmas and Easter are my days that you must keep. Don't even think about God's days, just forget them. Even though God gave them to you, I tell you they are not good for you, and should be thrown out. And if I can't get you to actually worship the sun, at least stick it in the Father's face. Keep the attributes of the solstice celebrations at the appropriate time of the winter solstice. And Christmas trees, with all their evergreen eternity, and burning logs, and don't forget the fertility of rabbits, lets, be sure and keep that symbolism in your holy days and their presence in the temple. Remember, that many pagans came to

God now calling themselves Christian, so how can this be wrong? See, paganism draws people to God.

"You will be doing it for god, because I am the true god, not Jesus, not the Father. Forget you know Christ wasn't born on December 25, I tell you it doesn't matter because the sun is reborn every year around that time.

"Now do as I say! Prove I am the god worthy of your obedience, accepting these demands, and by your actions, bow down before a tree and worship me!"

Tough words, very hard to write, especially as the sinner I am. I can tell you I daily pray for God's help to clean up my part of the temple. I am scared writing this book, realizing the hypocrite I am, at times. I daily fight lust, pornography, adultery, anger, drunkenness, selfishness, and cursing, to name a few (God knows). And I am thankful for my God-given precious wife who prompts me to show instead the fruits of the Spirit—love, grace, peace etc., and God knows I am trying (and I know it is only Christ and the Father in me that can actually make it happen by their grace).

Grace, mercy and peace from God the Father and from Jesus Christ, the Father's Son, will be with us in truth and love. (2 John 3, NIV)

But the fruit of the Spirit is love, joy, peace, patience, kindness, goodness, faithfulness, gentleness and self-control. Against such things there is no law. Those who belong to Christ Jesus have crucified the sinful nature with its passions and desires. Since we live by the Spirit, let us keep in step with the Spirit. (Galatians 5:22–25, NIV)

So it is my intent to **not** be arrogant or self-righteous in saying I do not keep pagan holy days and I think God hates them. I just humbly open my eyes to Gods words on these matters and try not to listen to man's or Satan's voice.

Speaking of voices, I sing with several groups at Christmas time and focus on the good, as we focus on the true salvation of Israel, the birth of the Son of God, Jesus. Members of those groups know I have no Christmas tree and that I am keeping to me a day of Purim at that time. Purim is an acceptable, manmade, God-respected day of celebrating the salvation of Israel. Mordecai and Queen Esther started it and gave gifts to each other as part of the original Purim celebration. They were also thankful for the salvation of Israel at that time. That is what I am borrowing from their actions. I present it as Purim to my God and show appreciation by giving gifts to others as we celebrate the salvation of Israel (Christ's birth, even though it is surely the wrong date).

I have been called a hypocrite for joining in with other Christians at that time. But I know their intent is to worship the true God, and God knows that I love and respect all the choir members. I just hope we will someday all call it Purim and, after realizing God's word, are willing to sweep the temple clean, as King Josiah did.

I hope it is clear I am not condemning you. The temple is what you are; it is just intended as an eye-opener to show us some of the unclean things we have all let in. So we are, you and me, the temple. We just need a little cleaning (for me a lot). Let's together listen to God and keep His desires.

God does not hate and very much loves all who intend to worship him, so I guess I am saying even though most Christians keep pagan days (or at least their attributes) and may thereby be at times 100 percent wrong, by the grace of God, He forgives and carries on, making us all 1,000 percent right. I present these things I call true because if you are like me, you will want to do as God desires. Please, let us together clean up the temple to God's liking. Like my elder brother said, "Get behind me, Satan, for we know the true Father, and we worship the true God in Jesus's name. If the words above are true, then, what you gonna do? Now that's up to you.

The Example of the Temple of God
What is the temple of God composed of?

*I did not see a temple in the city, because **the Lord God Almighty and the Lamb are its temple**. The city does not need the sun or the moon to shine on it, for **the glory of God gives it light, and the Lamb is its lamp**. The nations will walk by its light, and the kings of the earth will bring their splendor into it. On no day will its gates ever be shut, for there will be no night there. The glory and honor of the nations will be brought into it. Nothing impure will ever enter it, nor will anyone who does what is shameful or deceitful, but only those **whose names are written in the Lamb's book of life**. (Revelation 21:22–27, NIV)*

*What agreement is there between the temple of God and idols? **For we are the temple of the living God. As God has said: "I will live with them and walk among them, and I will be their God, and they will be my people."***
"Therefore come out from them
and be separate, says the Lord.
Touch no unclean thing,
and I will receive you." (2 Corinthians 6:16–17, NIV)

*Consequently, you are no longer foreigners and aliens, but fellow citizens with God's people **and members of God's household,** built on the foundation of the apostles and prophets, with **Christ Jesus himself as the chief cornerstone. In him the whole building is joined together and rises to become a holy temple in the Lord. And in him you too are being built together to** become a dwelling in which God lives by his Spirit. (Ephesians 2:19–22, NIV)*

Wow! The temple is the spiritually composed family of God, Father, Son, and bride! No third person spirit in the true temple of God.

I say again, beware of this third person (spirit) antichrist that appears as an angel of light at the temple, inserting himself as god seeking your worship. Just as he tried to deceive Christ, he tries to deceive us into believing he is God (he is not). The Trinity is NOT.

Concerning the coming of our Lord Jesus Christ and our being gathered to him, we ask you, brothers, not to become easily unsettled or alarmed by some prophecy, report or letter supposed to have come from us, saying that the day of the Lord has already come. **Don't let anyone deceive you in any way, for [that day will not come] until the rebellion occurs and the man of lawlessness is revealed,** *the man doomed to destruction.* **He will oppose and will exalt himself over everything that is called God or is worshiped, so that he sets himself up in God's temple, proclaiming himself to be God.** *Don't you remember that when I was with you I used to tell you these things? And now you know what is holding him back, so that he may be revealed at the proper time. For the secret power of lawlessness is already at work; but the one who now holds it back will continue to do so till he is taken out of the way. And then the lawless one will be revealed,* **whom the Lord Jesus will overthrow with the breath of his mouth** *and destroy by the splendor of his coming. The coming of the lawless one will be in accordance with the work of Satan displayed in all kinds of counterfeit miracles, signs and wonders, and in every sort of evil that deceives those who are perishing. They perish because they refused to love the truth and so be saved. For this reason* **God sends them a powerful delusion so that they will believe the lie.** *(2 Thessalonians 2:1–11, NIV)*

A final new thought on "What is there?" God is light, is spirit, is love, is our temple. Jesus is light, is spirit, is love, is our temple (John 4:24, John 4:8, 1 John 4:16, 1 John 1:5, 2 Corinthians 3:17, 18, Revelation 21:23). Nowhere do you find these attributes applied to a third individual. Only the Father and the Son are shown to be love and light and spirit. <u>The spirit is not a person, any more than love or light are persons.</u> These are attributes of, or the essence of, what God is made of.

> *I did not see a temple in the city, because* **the Lord God Almighty and the Lamb are its temple.** *The city does not need the sun or the moon to shine on it,* ==**for the glory of God gives it light, and the Lamb is its lamp.**== *The nations will walk by its light, and the kings of the earth will bring their splendor into it. (Revelation 21:22–24, NIV)*

If you believe in three persons of the Godhead, then perhaps you feel sorry for the "spirit person" for it is missing again and not shown as part of the temple. Of course, if God is made of Holy Spirit, then when the Father person and the Son person are present, so is IT because God is made of Spirit? Is God really a temple literally? Is God literally light? Before creation was God, so everything that exists, wouldn't it be some form of his essence? Where else would it come from?

> *The Glory of Zion*
> ***"Arise, shine, for your light has come,***
> *and **the glory of the LORD rises upon you**.*
> *See, darkness covers the earth*
> *and thick darkness is over the peoples,*
> *but the LORD rises upon you*
> *and his glory appears over you.*
> *Nations will come to your light,*
> *and kings to the brightness of your dawn.*
> *"Lift up your eyes and look about you:*

All assemble and come to you;
your sons come from afar,
and your daughters are carried on the arm.
Then you will look and be radiant,
your heart will throb and swell with joy;
the wealth on the seas will be brought to you,
to you the riches of the nations will come.

Surely the islands look to me;
in the lead are the ships of Tarshish,
bringing your sons from afar,
with their silver and gold,
to the honor of the LORD your God,
the Holy One of Israel,
for he has endowed you with splendor.

The sun will no more be your light by day,
nor will the brightness of the moon shine on you,
for the LORD will be your everlasting light,
and your God will be your glory.
Your sun will never set again,
and your moon will wane no more;
the LORD will be your everlasting light,
and your days of sorrow will end.
(Isaiah 60:1–5, 9, 19–20 (NIV)

So to end of this chapter, I ask, "Knock, knock! Who's there?" For me, the answer is the Father and the Son (and eventually, bride). But if I ask, "Knock, knock! What is there?" For me, the answer is the persons Father, and Son (and eventually, bride) made up of and sealed in the essence of Holy Spirit (like living water). And the Father and Son are love, light, and spirit (and we eventually shall be like them in a spiritual oneness).

CHAPTER 6: MY CREDENTIALS

I was a lay minister (local elder) with the Worldwide Church of God, in Calgary, Alberta, Canada. I was born and raised in Calgary and now live in Cochrane, Alberta, just west of Calgary, closer to the beloved Rocky Mountains.

My earliest memories of God came from a neighbour who was kind enough to take me into her home and play music and sing about God to me. She was a member of the Emmanuel Church on First Street and Seventeenth Avenue Southwest Calgary. I think this was God putting his hand on me, and I thank that lady very much for being His servant.

But before that, when I was about five years old, I can remember a special day. It was a beautiful warm summers day, and I had just successfully tied my shoes for the first time and proudly stepped out into our street. I remember looking up and, feeling deeply in my heart, saying "Thank you for making me. I am so glad to be here." I didn't know I was speaking to God. I was just happy and knew something was responsible for this wonderfulness. I think that fragile seed somehow fell on good soil in a place and time where good soil was desperately scarce—the '60s.

I would describe myself as a child on my own. In some ways, it was a much safer place and time. I remember leaving the house, wondering off for hours and miles with no concerns and nobody with me. My loving parents were not ready for the '60s and kids. I was left to my own and became a bit of a monster—drink, drugs, and rock and roll. Satan was killing the garden. To emphasize how different times were then, I find it hard to believe, but I joined little league baseball and my parents didn't know. At least I don't think so. They certainly never took me to a game or ever came to one. Strange, huh?

One Sunday, I asked for a dime because I just wanted to go to church. I went to the Knox Presbyterian Church on the corner of Thirty-Fifth Avenue and Thirty-Seventh Street. I found a spot where the sun shone through some beautiful-colored glass and sat there with respect for this loving God that I didn't know. There were no other people around me, and I was happily kicking my feet, not listening to a word being said. I am sure I was driving the poor preacher nuts. He kept looking up at me. He was not a happy preacher. But I didn't care. I was somehow fulfilling a need to please God just by being there.

Years later, early one morning, I and several of my teenage friends sat drunk, and drinking outside that same church. I drove them there because there was something I remembered, something special inside those doors, a closeness to something good.

By this time, I had my indoctrination into the loftiness of evolution, had bought it hook, line, and sinker, except I couldn't help at least hoping, there is something more to life. ***We are not just dust in the wind, having come from nothing, slowly.*** God somehow still had a line to me. But what happened? How could it come to this terrible state?

In 1964, the radio was on, and I heard something captivating. It was the Pied Piper. No, it was the Beatles, but it was a Pied Piper effect. Wow, it was wonderful, and it had an immediate hold on me—music. It was precious. Add to that junior high with grade 7 girls, and let me out of the house now! Party time.

God's angels are busy, aren't they? Thank you, God, and the angels.

The result could be expected. I ran away from home at about thirteen years of age but was lucky my parents let me back. I attempted suicide at that same age, and my parents never came to the hospital. Dad never ever talked about it. Not once. Mom was hurt but was only cold and mad about it. Perhaps the psychologists told them to be that way? A child on his own would surely come to this. I was a bad ass. I was kicked out of both the high schools I attended. So with a grade 10 education, I got myself a Norton 750 Commando motorcycle and joined society's workforce. I shovelled asphalt for the city of Calgary.

As a teenager, I believe God had not given up on me and actually saved my physical life. One time. I was a passenger in a Volkswagen beetle, again, drunk (at least no LSD in us). We left Channel 4 Hill and drove south on Sarcee Trail. I was in the back on the right side, and I remember the driver saying, "Shit! Hold on!" We hit a curb, doing about eighty miles an hour, and I tilted to the right.

Then there was a dreamy floating feeling. I was semiconsciously sliding on the grass, down the divide in the road. Then after stopping, I opened my eyes and looked up, and someone from a passing car came up to me and, looking down on me, said, "Wow! You are alive!" He had witnessed the car roll over four times. I had been thrown out the tiny backseat side window (that only pops open about four inches) and was lying safe but in shock. I had cuts on the back of my knees and hurt, but I was able to get up and try to hide the beer bottles lying all around me. I have faith that I was protected that night.

Another time, I was on my motorcycle, and this one had a lasting effect on me.

On Ninth Avenue southwest, heading east doing about 45 mph., I approached Fifth Street, and to my horror, a red mustang left a stop sign, pulled out in front of me, and deliberately stopped. I believe he was trying to kill me. I felt the front wheel sink into the side door of the car, and my knees scraped the roof, as I flew up into a 360-degree flip and landed on my butt. The picture of the Robin Hood Flour Mill is still in my brain, looking strangely upside down, as I did my flight. I don't remember them tearing that old building down, but for me it will never leave. Forty-four years later, at the last job I had, I was working on the thirty-ninth floor of the Eight Avenue Place tower and could look down on the accident site at Fifth Street and Ninth Avenue (now an underpass).

My lower back has never forgiven that driver, who was charged with dangerous driving. The police told me witnesses spoke up for me. I never had the brains to sue the guy, as my bike was totalled and my lower back has a nasty curve down low that gets me these days.

I have faith God saved me that day and that God has since gone to the effort to remind me of it in His wonderful personal ways. This reminder happened almost every day I entered work with Whitecap Resources in 2016. The building is at the very same corner where my accident (attempted murder?) happened.

At almost the exact spot my bike collided, there is now a low-hanging branch from a tree that the Cochrane Southland Commuter bus I used brushed as it passed my building before dropping me off at Gulf Canada Square (the place where the old Robin Hood flour mill had once been). The brush hit the bus daily, pretty much exactly where I previously had my brush with death all those years ago. I told the driver my strange story. He is a friend of mine named Tom and was the best driver I ever had at Southland Transportation.

It was the year before that when God moved in and steered me straight. I had a new girlfriend who was religious. At least her dad was. I found this out when I tried to get her to take drugs (MDA, the '60 love drug). She told me her dad says that "drugs are from the evil one and will harm us." She didn't want to disappoint her dad.

I was perplexed because of my haunting past that was inwardly conflicting with my now much more intellectual understanding that our creator is evolution, and chance. But I rode around for a week, fighting this desire to do drugs together. Eventually, I gave in and convinced her to take these drugs so we could have sex together in my parents' basement. I used that brilliant argument that if there was a God who wanted us in heaven, he would have simply put us there. "So don't listen to your dad's foolish ideas." But I knew a different feeling was tugging at my heart.

It was about an hour after we had taken these speed-like MDA, but things were different this time. Instead of being sped up and active, this drug had shut us right down. I instinctively realized we were in trouble, as neither of us was moving. I have no idea how much danger we were in, but it was enough for me to rethink our situation.

I was scared, and I prayed for the first time in my life. I said, "I am sorry for what I have gotten your daughter into. She didn't want to do this. Please don't blame her, blame me. Please heal her, and I don't care what happens to me." We both then immediately sat up straight and sober, and she looked at me and said, "Jesus was just here." She had not heard my previous silent prayer. I was no longer a drug addict from that moment on. God healed us both by His tremendous grace.

I asked her to take me to her father, so she did. I subsequently joined God's boot camp, the Worldwide Church of God under Herbert W. Armstrong, with the World Tomorrow and Plain Truth booklets to protect me. My friends were horrified and began to laughingly call me "the preacher." Yes they laughed, but they believed me.

There were several other substantial miracles that occurred later that I don't think will improve my reasons for mentioning this all. I realize now that this is a story about just one seed from the sower that hit pay dirt (I hope) and, miraculously, against all the things Satan can throw at seemingly defenceless children, still lives, growing in grace and knowledge. It is the reason I dare to write this book. I want to share the good gift and have it grow to many seeds, as per Gods will. His word, his seed, will not return to Him empty. And I will stand up for God and publicly praise Him in this way, to help other prodigal souls who have never heard the wonderful news about God and His purpose for them.

I regret nothing of those WCG church days, not even tithing up to over 30 percent of my gross income in third tithe year. We were 100 percent wrong about certain things, but at the same time, 1,000 percent right in our hearts.

I was already on the road to recovery when that motorcycle accident happened while I was heading to school that I had gone back to. As teenagers, we instinctively knew that doing things like sniffing glue, getting roaring drunk, and doing drugs was bad for our brains. But we encouraged each other into doing this because as nothing but dust in the evolutionary wind, we didn't care if we destroyed ourselves, at least we could maybe salvage some fun out of it. But now with new hope, I was certain God wanted me to get an education, so I went back. One day, while I was shovelling asphalt in Victoria Park, I looked up and remember thinking God would want me to get educated.

Shortly thereafter, I heard a radio announcement about a PEP program (priority employment program) at the Alberta Vocational Center (now Bow Valley College), where I got paid unemployment insurance to attend high school matriculation. You had to take some tests to get in, and I remember a Mr. Birdhall calling me about my results and surprising me by saying I could be a doctor if I wanted to. It all just fell into place. God made it happen. Of that I am certain.

And now I am a survivor of the '60s and '70s. And feel I have the answer to the question of my generation, which I have a driving heart to share. Perhaps my favorite rock and roll group of all time said it all so well. At least I think it sums me up pretty much perfectly. I had forgotten my God, but he had not forgotten me! From the Who, the song is "**Who are You?**" Below is an explanation of the words, and below that are the words my heart echoes. I am thankful for the answer so I am sharing this with you.

Through the words of this book, I hope we all know who God is and who we are, Israel. **And the promise is our graceful place in outer space.**

Figure 47. The Prayer of a Desperate Man (http://www.songfacts.com/detail.php?id=2394)

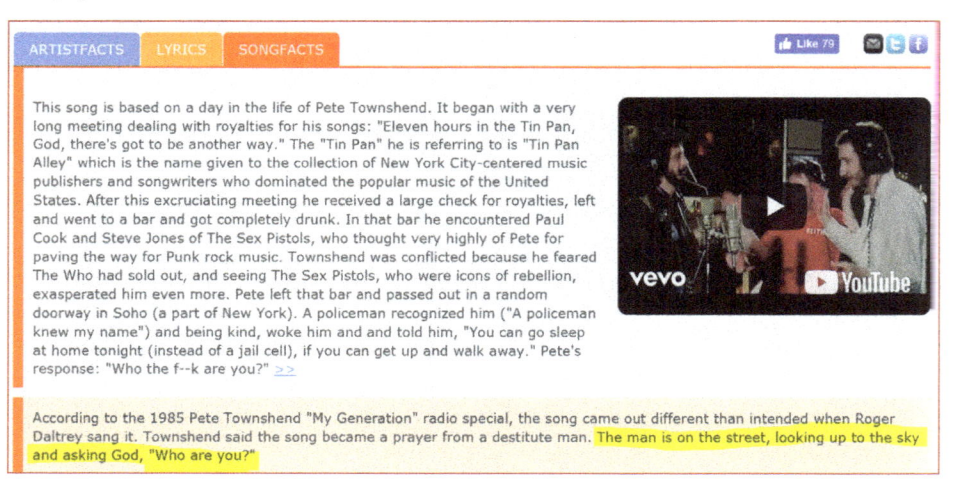

Source: http://www.songfacts.com/detail.php?id=2394

SPACE THE TRUE FRONTIER!

"Who Are You"

Who are you?
Who, who, who, who?
Who are you?
Who, who, who, who?
Who are you?
Who, who, who, who?
Who are you?
Who, who, who, who?

I woke up in a Soho doorway
A policeman knew my name
He said "You can go sleep at home tonight
If you can get up and walk away"

I staggered back to the underground
And the breeze blew back my hair
I remember throwin' punches around
And preachin' from my chair

[chorus:]
Well, who are you? (Who are you? Who, who, who, who?)
I really wanna know (Who are you? Who, who, who, who?)
Tell me, who are you? (Who are you? Who, who, who, who?)
'Cause I really wanna know (Who are you? Who, who, who, who?)

I took the tube back out of town
Back to the Rollin' Pin
I felt a little like a dying clown
With a streak of Rin Tin Tin

I stretched back and I hiccupped
And looked back on my busy day
Eleven hours in the Tin Pan
God, there's got to be another way

Who are you?
Ooh wa ooh wa ooh wa ooh wa ...

Who are you?
Who, who, who, who?
Who are you?
Who, who, who, who?
Who are you?
Who, who, who, who?
Who are you?
Who, who, who, who?

[chorus]

I know there's a place you walked
Where love falls from the trees
My heart is like a broken cup
I only feel right on my knees

I spit out like a sewer hole
Yet still recieve your kiss
How can I measure up to anyone now
After such a love as this?

Figure 48. Who Are You – To the Great I AM (The Who)

And so who am I? I am nothing, except some dust that God's breath moved. But that is all you need.

> *For the preaching of the cross is to them that perish foolishness; but unto us which are saved it is the power of God.*
> *For it is written, I will destroy the wisdom of the wise, and will bring to nothing the understanding of the prudent.*
> *Where is the wise? where is the scribe? where is the disputer of this world? hath not God made foolish the wisdom of this world?*
> *For after that in the wisdom of God the world by wisdom knew not God, it pleased God by the foolishness of preaching to save them that believe.*
> *For the Jews require a sign, and the Greeks seek after wisdom:*
> *But we preach Christ crucified, unto the Jews a stumblingblock, and unto the Greeks foolishness;*
> *But unto them which are called, both Jews and Greeks, Christ the power of God, and the wisdom of God.*
> *Because the foolishness of God is wiser than men; and the weakness of God is stronger than men.*
> *For ye see your calling, brethren, how that not many wise men after the flesh, not many mighty, not many noble, are called:*
> *==But God hath chosen the foolish things of the world to confound the wise; and God hath chosen the weak things of the world to confound the things which are mighty;==*
> *And base things of the world, and things which are despised, hath God chosen, yea, and things which are not, to bring to nought things that are:*
> *That no flesh should glory in his presence.*
> *But of him are ye in Christ Jesus, who of God is made unto us wisdom, and righteousness, and sanctification, and redemption:*

SPACE THE TRUE FRONTIER!

That, according as it is written, He that glorieth, let him glory in the Lord. (1 Corinthians 1:18–31, KJV)

Again, I am dust with just a single breath added, and that is enough to serve our God by His strength, by his righteousness. *Genesis 2:7 (NIV): "The LORD God formed the man from the dust of the ground and breathed into his nostrils the breath of life, and the man became a living being."*

Thank you, God, **Lord Jesus will overthrow with the <u>breath of his mouth.</u>**

EPILOGUE

Our future is sharing the existence of the glory of Christ with the Father in the glorious new heavens and new earth. Our place is outer space. Always has been.

God the Father and His Son Jesus are the temple. We are part of that same temple if we have Christ and the Father in us.

> *And I saw no temple therein: **for the Lord God Almighty and the Lamb are the temple** of it. And the city had no need of the sun, neither of the moon, to shine in it: for the **glory of God did lighten it,** and **the Lamb is the light thereof.** (Revelation 21:22–23, KJV)*

> *Don't you know that **you yourselves are God's temple and that God's Spirit lives in you?** If anyone destroys God's temple, God will destroy him; for God's temple is sacred, **and you are that temple**.) (1 Corinthians 3:16–17, NIV)*

And in the New World Translation of Romans 8:9–11, the Father and Jesus spiritually dwell in and comfort us, like a river of living water in us, making it so we are included as part of the true temple of God.

> 9 However, you are in harmony, not with the flesh, but with the spirit, + if God's spirit truly dwells in you. But if anyone does not have Christ's spirit, this person does not belong to him. 10 But if Christ is in union with you, + the body is dead because of sin, but the spirit is life because of righteousness. 11 If, now, the spirit of him who raised up Jesus from the dead dwells in you, the one who raised up Christ Jesus from the dead + will also make your mortal bodies alive + through his spirit that resides in you.

Figure 49. Christ and the Father, are the combined-Spirit Comforter, Source JW Bible

Satan denies that temple because he denies the Father and the Son. He does not deny himself, the third person around the temple of God. Satan says he is part of that temple, and I deny that by the breath of God that is in me, and thereby God reveals the great deceiver that fools the whole world, acting as an angel of light that he once was so long ago.
The ex-morning star, now Satan the antichrist:

> *He will oppose and will exalt himself over everything that is called God or is worshiped, so that **he sets himself up in God's temple, proclaiming himself to be God**. (2 Thessalonians 2:4, NIV)*

> *But **you have an anointing from the Holy One**, and all of you know the truth. I do not write to you because you do not know the truth, but because you do know it and because no lie comes from the truth. Who is the liar? It is the man who denies that Jesus is the Christ. **Such a man is the antichrist—he denies the Father and the Son**. No one who denies the Son has the Father; whoever acknowledges the Son has the Father also. (1 John 2:20–23, NIV)*

If there were three personalities in the Godhead, why didn't Satan deny the third? Most ignore this question and refuse to

communicate as Bereans (Acts 17:11). When people didn't answer Christ, often it was because they didn't want the answer—or the truth.

Whose voice do you listen to? *See that what you have heard from the beginning remains in you. If it does, you also will **remain in the Son and in the Father**. And this is what he promised us—even eternal life" (1 John 2:24–25).*

The powers against us, false religion, Satan, and faith in evolution, have fought hard to the point mankind has almost absolutely forgotten the promises and intent of God. Satan will soon be stopped from jamming the message that the sower has broadcast!

God has never forgotten us, and the promise to share in oneness with the Father and the Son in the glorious universe as his spiritual Israel (family). That which He has sown shall He reap. This will ALWAYS BE. NO VOICE WILL SILENCE IT. And He is coming soon to make it happen! I believe in our great future—the sons of God in spirit life go out to dress the universe.

Space the true frontier of eternity! ==Just imagine what it will become.==

> *I consider that our present sufferings are not worth comparing with the glory that will be revealed in us. The creation waits in eager expectation for the sons of God to be revealed. For the creation was subjected to frustration, not by its own choice, but by the will of the one who subjected it, in hope that the creation itself will be liberated from its bondage to decay and brought into the glorious freedom of the children of God. (Romans 8:18–21)*

> *The Spirit and the bride say, "Come!" And let him who hears say, "Come!" Whoever is thirsty, let him come; and whoever wishes, let him take the free gift of the water of life. (Revelation 22:17 NIV)*

I have taken a lot of time to bring that realization to you because I feel God wanted me too. I am not doing it to gain any fame; I am trying only to be a humble servant who is delighted to tell people about a long-lost inheritance they have, especially those that are not Christians yet, those who will come to work for God at the last part of the day and get paid the same. Yeah for them.

Before Christ comes back, this son of perdition will be revealed, the deception opened to our eyes, and perhaps I (being just one little breath of God) am the servant He has used.

> *Blessed is the God and Father of our **Lord Jesus Christ**, who, according to the abundance of His kindness did beget us again to a living hope, through the rising again of Jesus Christ out of the dead, to **an inheritance incorruptible, and undefiled, and unfading, reserved in the heavens for you**, who, in the power of God are being guarded, through faith, unto salvation, ready to be revealed in the last time.* (1 Peter 1:3–5, YLT)

I pray you are good soil for Gods Word, please pass it on to anyone thirsty around you.

And I hope from now on, for all of us who are God's children, grace will be enough, and we will not be forced to adopt the deceptions of the false god-Satan. Let us all say to the one who dared to change days, and times, and to insert himself as our father, 'Get thou behind me Satan', for we worship the Lord our God, and Him only:

Think of that inheritance we have.

It is better than a billion dollars, if you truly believe it.

If you were to find the ark of the covenant, and astonishingly find the actual jar of manna within it, I hope you realize that amazing historical treasure, would be paled to insignificance, compared to you picking up a Bible. For that is the true manna, Christ the Word, from heaven.

And I hope next time you look up at the stars at night, you will know in your heart, it's not just a lot of empty space out there, … that's our place…that's home!

In Jesus name, amen.

CREDITS

A friend of mine named Ken Trout was the first to share these thoughts about the deceiver being the third person in the false image of the Trinity Godhead. I want it to be known that I thank him.

George McIntosh, Wayne Vials, Eric Rasmussen, and Ken Trout stood up in 2007 against a lying voice and sheltered ourselves from a flood of false spirit. We left that church, as it turned to the voice of the Trinity and to the voice that changed seasons and times.

Today I have tried to provide a God-sanctioned rebuttal to that abomination, ten years in the making, hopefully truly inspired by God. At the same time, we needed grace and to grow more in it. There are many that God is working with.

I give credit to Mr. Herbert W. Armstrong, who had an undeniable heart for God and shared his hope with me. He never said he was perfect, but he stood up for God at all costs, as I hope do I.

Some scientists like to think of themselves as the top of the evolutionary chain in the universe. Little gods, I guess, who desire our worship too. They who deny that entropy requires intelligence to defeat and so, by default, deny the ultimate windup (big bang) would require ultimate power and intelligence to make happen. I deny them.

Scientists, those who speak of theories as facts, are they themselves revealed as foolishness in the word of God. Satan has thrown everything at us to hide our destiny and defeat God's purpose.

False religion has appeared as enlightenment. Science with evolution, and its indoctrinations from our childhood schools, says we came from nothing slowly. But the word of God, which cannot be destroyed, tells us the truth—God is our Creator. So I credit this Creator, God, who opens our eyes and ears by His mighty Spirit presence living in us as Father and Son. Thank you, God.

I can promise you all, I share this in good conscience and with a good heart. To the greater body of Christ's grace, I love you all, so let's get cleaning our part of the temple; it needs some work. This is truth that I believe God has given me (let us all grow in grace and knowledge), and God knows I share it in honestly.

I will stand up for God at all costs in this, and if He shows me different, I will change to God's will.

Would you?

APPENDICES

1 Entanglement at a distance.

Meaning of entanglement

An entangled system is defined to be one whose quantum state cannot be factored as a product of states of its local constituents; that is to say, they are not individual particles but are an inseparable whole. In entanglement, one constituent cannot be fully described without considering the other(s). Note that the state of a composite system is always expressible as a *sum*, or superposition, of products of states of local constituents; it is entangled if this sum necessarily has more than one term.

Quantum systems can become entangled through various types of interactions. For some ways in which entanglement may be achieved for experimental purposes, see the section below on methods. Entanglement is broken when the entangled particles decohere through interaction with the environment; for example, when a measurement is made.

As an example of entanglement: a subatomic particle decays into an entangled pair of other particles. The decay events obey the various conservation laws, and as a result, the measurement outcomes of one daughter particle must be highly correlated with the measurement outcomes of the other daughter particle (so that the total momenta, angular momenta, energy, and so forth remains roughly the same before and after this process). For instance, a spin-zero particle could decay into a pair of spin-½ particles. Since the total spin before and after this decay must be zero (conservation of angular momentum), whenever the first particle is measured to be spin up on some axis, the other, when measured on the same axis, is always found to be spin down. (This is called the *spin anti-correlated* case; and if the prior probabilities for measuring each spin are equal, the pair is said to be in the singlet state.)

The special property of entanglement can be better observed if we separate the said two particles. Let's put one of them in the White House in Washington and the other in UC Berkeley (think about this as a thought experiment, not an actual one). Now, if we measure a particular characteristic of one of these particles (say, for example, spin), get a result, and then measure the other particle using the same criterion (spin along the same axis), we find that the result of the measurement of the second particle will match (in a complementary sense) the result of the measurement of the first particle, in that they will be opposite in their values.

The above result may or may not be perceived as surprising. A classical system would display the same property, and a hidden variable theory (see below) would certainly be *required* to do so, based on conservation of angular momentum in classical and quantum mechanics alike. The difference is that a classical system has definite values for all the observables all along, while the quantum system does not. In a sense to be discussed below, the quantum system considered here seems to *acquire* a probability distribution for the outcome of a measurement of the spin along *any* axis of the *other* particle upon measurement of the *first* particle. This probability distribution is in general *different* from what it would be *without* measurement of the first particle. This may certainly be perceived as surprising in the case of spatially separated entangled particles.

Paradox

The paradox is that a measurement made on either of the particles apparently collapses the state of the entire entangled system—and does so instantaneously, before any information about the measurement result could have been communicated to the other particle (assuming that information cannot travel faster than light) and hence assured the "proper" outcome of the measurement of the other part of the entangled pair. In the Copenhagen interpretation, the result of a spin measurement on one of the particles is a collapse into a state in which each particle has a definite spin (either up or down) along the axis of measurement. The outcome is taken to be random, with each possibility having a probability of 50%. However, if both spins are measured along the same axis, they are found to be anti-correlated. **This means that the random outcome of the measurement made on one particle seems to have been transmitted to the other, so that it can make the "right choice" when it too is measured.**

The distance and timing of the measurements can be chosen so as to make the interval between the two measurements spacelike, hence, any causal effect connecting the events would have to travel faster than light. According to the principles of special relativity, it is not possible for any information to travel between two such measuring events. It is not even possible to say which of the measurements came first. For two spacelike separated events x_1 and x_2 there are inertial frames in which x_1 is first and others in which x_2 is first. Therefore, the correlation between the two measurements cannot be explained as one measurement determining the other: different observers would disagree about the role of cause and effect. (https://en.wikipedia.org/wiki/Quantum_entanglement)

2. Double-slit experiment

Scientists are (and should be) perplexed by several realities about this experiment. If I understand it right, they are guessing astounding possible explanations, like string theory with parallel universes, things much more far stretched than the simple statement "God is light," and that's how it knows because it is God (spirit in a different state; frozen God). *"This then is the message which we have heard of him, and declare unto you, that* **God is light***, and in him is no darkness at all" (1 John 1:5, KJV).*

The other astounding observation is that observing the photon changes the whole outcome! I am wondering if there is a parallel between qubit superposition (wave interference) and base state (no wave interference caused by peeking). Is peeking like grabbing a flipping coin, causing it to go back to the base state? As stated below, when there are two slits, if you attempt to observe which slit the light goes through, you don't get refraction, and it is as though there is only one slit (amazing). What could possibly let the photon know you are watching? I think our amazing God is everywhere, and everything and everywhen. So God (who is light) is right every time. Again, this is my suspicion only, and it doesn't change the reality that something is informing the photons of the differences and, to do this, must be faster than light to get the change to be effective in time.)

So you choose—parallel universes or trust that God is light and therefore instantly knows at the source. Anyway, it is fun stuff. Our God and universe is totally incredible.

https://ca.answers.yahoo.com/question/index?qid=20130522011057AARI9Rn

https://www.youtube.com/watch?v=LW6Mq352f0E

Society & Culture > Religion & Spirituality   Next >

In the double slit experiment, does God know which slit the photon goes through?

Here is an explanation of the double slit experiment: http://www.youtube.com/watch?v=LW6Mq352f...

In short it is simple. If you shine light on a double slit, it produces a diffraction pattern (i.e. light behaves as a wave and goes through BOTH slits and interferes with itself). The problem is two-fold:

1) If you reduce the light source such that only one photon goes through at a time, you will indeed see one photon hit the screen at a time. However if you let this go for a long time, you will see a diffraction pattern emerge. The problem here is that light is coming in as a particle (it's only hitting one part of the screen per photon emitted through the double slit).

2) If you attempt to observe which slit the light goes through, the diffraction pattern disappears and, instead, you get a pattern consistent with particles.

So my question is: does God observe which slit the photon goes through or does God not KNOW which slit it goes through (unless humans observe it)?

SPACE THE TRUE FRONTIER!

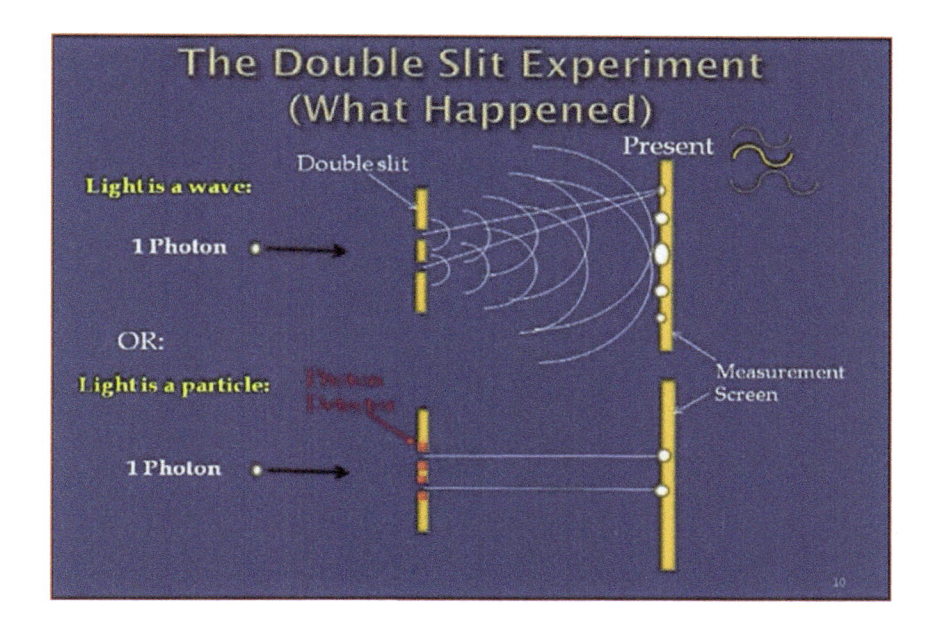

Source:
https://www.bing.com/videos/search?q=double+slit+experiment&view=detail&mid=B3B630C0EDD09988BA4EB3B630C0EDD09988BA4E&FORM=VIRE

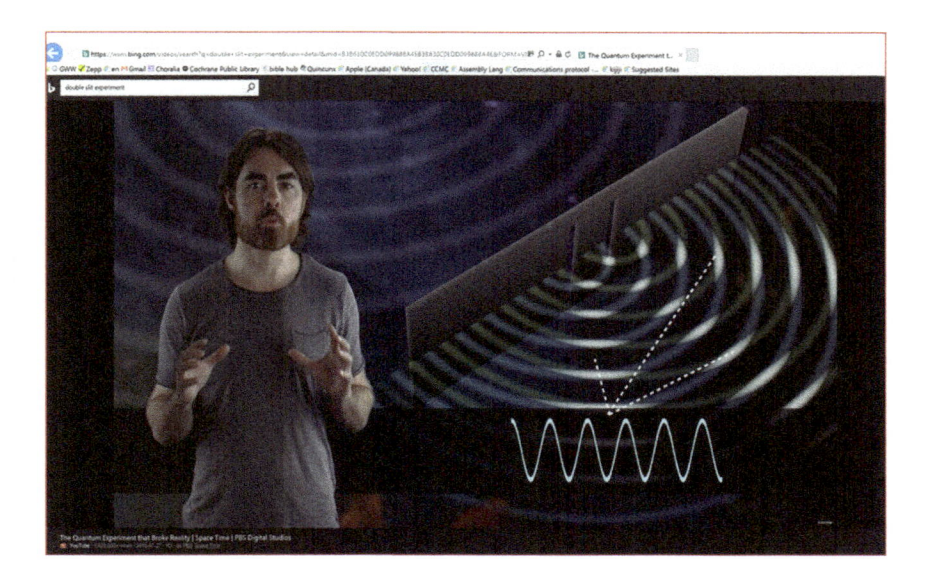

3. The universe is math (to some scientists).

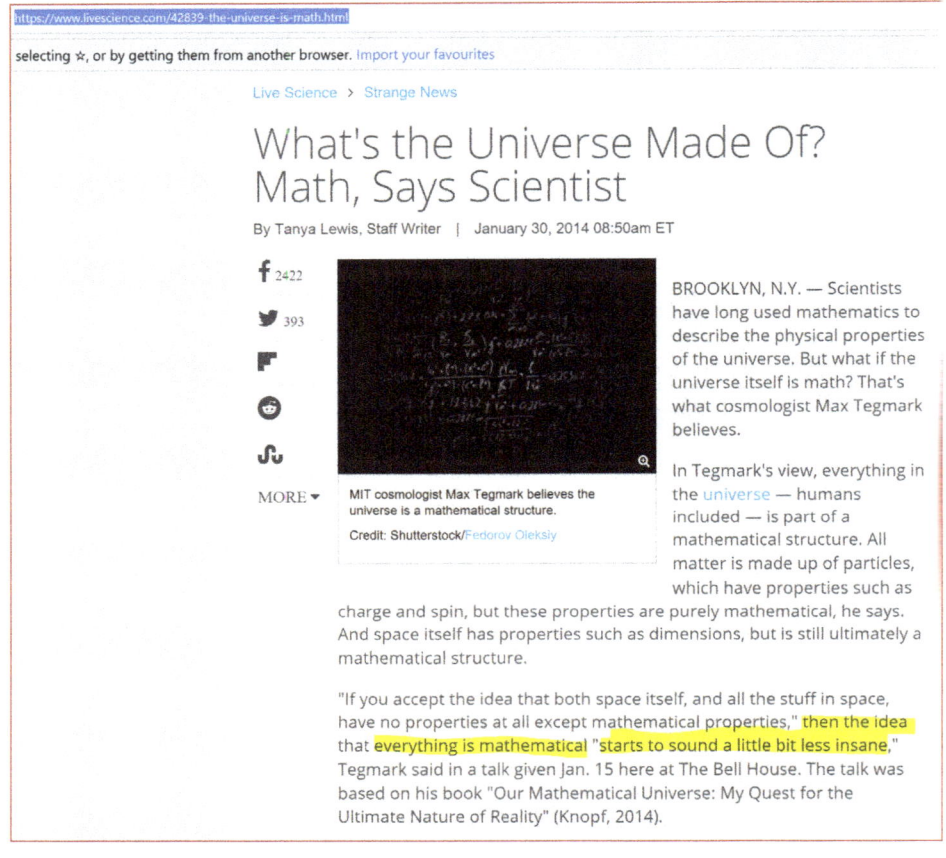

> **What's the Universe Made Of? Math, Says Scientist**
>
> By Tanya Lewis, Staff Writer | January 30, 2014 08:50am ET
>
> BROOKLYN, N.Y. — Scientists have long used mathematics to describe the physical properties of the universe. But what if the universe itself is math? That's what cosmologist Max Tegmark believes.
>
> In Tegmark's view, everything in the universe — humans included — is part of a mathematical structure. All matter is made up of particles, which have properties such as charge and spin, but these properties are purely mathematical, he says. And space itself has properties such as dimensions, but is still ultimately a mathematical structure.
>
> "If you accept the idea that both space itself, and all the stuff in space, have no properties at all except mathematical properties," then the idea that everything is mathematical "starts to sound a little bit less insane," Tegmark said in a talk given Jan. 15 here at The Bell House. The talk was based on his book "Our Mathematical Universe: My Quest for the Ultimate Nature of Reality" (Knopf, 2014).
>
> *MIT cosmologist Max Tegmark believes the universe is a mathematical structure.*
> *Credit: Shutterstock/Fedorov Oleksiy*

Source: https://www.livescience.com/42839-the-universe-is-math.html

4. Invisible spirit becomes matter?

Hebrews 11:3 (NIV): "By faith we understand that the universe was formed at God's command, so that <u>what is seen</u> was not made out of what was visible."

Science agrees that scripture is true, for the latest physics states all matter requires a Higgs boson field. And science says that before that Higgs boson field made it happen, there was no mass and no matter (nothing physical). This field cannot be seen but is supposedly proven to exist because they have seen the Higgs boson particle since 2012 (also known for that reason as the "God particle"). I say, out of invisible spirit (the true God field) came the physical universe (matter). So perhaps my idea is not such a great stretch of the imagination after all!

I think God's Word = Spirit.

> *But they deliberately forget that long ago <u>by God's word the heavens existed and the earth was formed out of water and by water.</u> (2 Peter 3:5, NIV)*

> *But in these last days he has spoken to us by his Son, whom he appointed heir of all things, and <u>through whom he made the universe</u>. The Son is the radiance of God's glory and the exact representation of his being, <u>sustaining all things by his powerful word</u>. After he had provided purification for sins, he sat down at the right hand of the Majesty in heaven. (Hebrews 1:2–3, NIV)*

> *Take the helmet of salvation <u>and the sword of the Spirit, which is the word of God</u>. And pray in the Spirit on all occasions with all kinds of prayers and requests. With this in mind, be alert and always keep on praying for all the saints. (Ephesians 6:17–18, NIV)*

> *The Spirit gives life; the flesh counts for nothing. <u>The words I have spoken to you are spirit</u> and they are life. (John 6:63, NIV)*

<u>If I am correct in my thinking, according to the above scriptures, the Word of God (which is Jesus and is Spirit) became matter (the universe) and is therefore Spirit in a solid matter form. And this invisible spirit (or according to science, the Higgs Boson field), which cannot be seen, became that which can be seen (the universe).</u> And from science, what a tremendous confirmation of the Bible. Science once again has confirmed ancient scripture, the word of God.

Please stop and think how marvellous it is that what the Bible revealed <u>anciently</u> could not have humanly been known in its fullness without the incredible knowledge of our present times (science).

I am also referring to what for me may be other confirmations, like that the heavens exist above the circle of the earth (Isaiah 40:22) and that the tiniest particles are really waves, perhaps like the essence of God (His Spirit) that fluttered (waved) upon the waters and brought things into existence. Special relativity from Einstein shows that time is relative and not absolute. I propose that is how God knows the end from the beginning and perhaps that is how a thousand years is like a day to Him.)

> *And this one thing let not be unobserved by you, beloved, that one day with the Lord is as a thousand years, <u>and a thousand years as one day</u>. (2 Peter 3:8, YLT)*

> *He who is <u>sitting on the circle of the earth</u>, And its inhabitants are as grasshoppers, He who is stretching out as a thin thing the heavens, And spreadeth them as a tent to dwell in. (Isaiah 40:22, YLT)*

> *<u>God sits high above the round ball of earth</u>.*
> *The people look like mere ants.*
> *He stretches out the skies like a canvas—*
> *yes, like a tent canvas to live under. (Isaiah 40:22, MSG)*

> *The earth hath existed waste and void, and darkness is on the face of the deep, and the Spirit of God <u>fluttering</u> on the face of the waters. (Genesis 1:2, YLT)*

SPACE THE TRUE FRONTIER!

[Screenshot of Interlinear Scripture Analyzer showing Genesis 1:2 with Hebrew text and interlinear translation: "And the earth was without form, and void; and darkness [was] upon the face of the deep. And the Spirit of God moved upon the face of the waters." The concordant view shows H7363 with renderings "he-is-vibrating", "vibrating", and "they-vibrate" highlighted.]

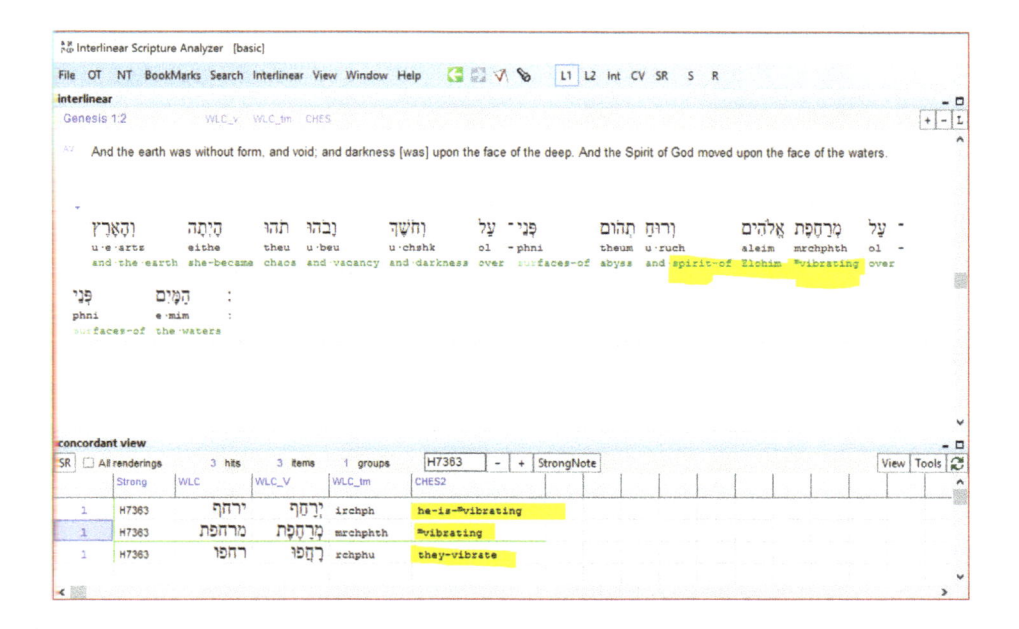

"This then is the message which we have heard of him, and declare unto you, that <u>God is light</u>, and in him is no darkness at all" (1 John 1:5, KJV). Yes, and what is light, science can tell us much more today. It is tremendous what they have found out, but it is still mysterious. Einstein described light as photon particles, but most of science now agrees that light is really a wave that vibrates.

Some scientists say light is both a particle and a quantum wave (or packet of particles) at the same time. The funniest observation about light is that they say it is a particle when you look at it, but if you look away, it is a wave! That when you observe it, you affect a change on it. The famous experiment called the double slit is totally fascinating, and I have provided some info on that for you (see Appendix #2).

The point I want to make is, we should be delighted with these scientific insights. The Bible is nothing if it is not true and in line with the physical reality of the universe. Science has been backing up the Bible for centuries, and we Christians should not be afraid of their discoveries.

True science has great information, and science has corrected the ignorance of some theologians (but not the Bible) throughout history. For example, the earth is not the center of the solar system. The earth is not flat.

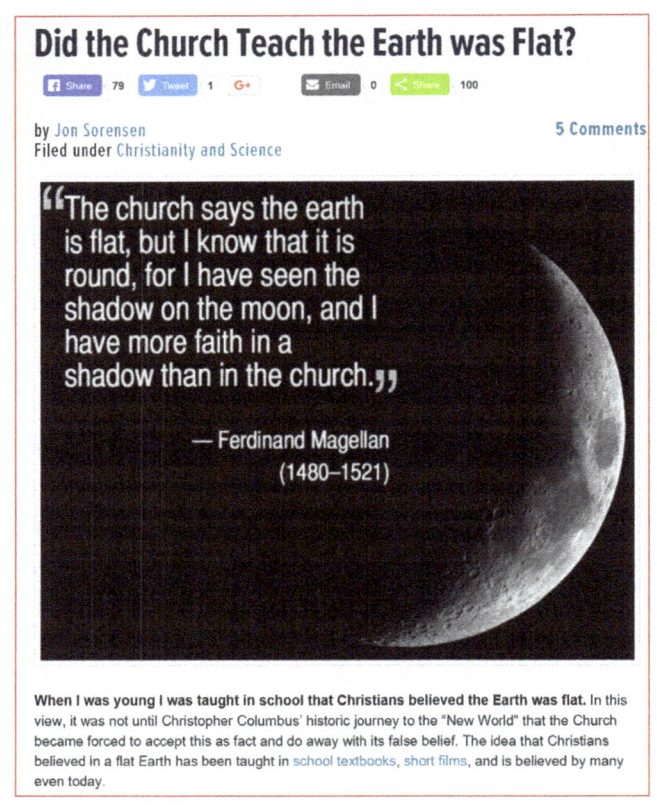

> **Did the Church Teach the Earth was Flat?**
>
> by Jon Sorensen
> Filed under Christianity and Science
> 5 Comments
>
> "The church says the earth is flat, but I know that it is round, for I have seen the shadow on the moon, and I have more faith in a shadow than in the church."
>
> — Ferdinand Magellan (1480–1521)
>
> When I was young I was taught in school that Christians believed the Earth was flat. In this view, it was not until Christopher Columbus' historic journey to the "New World" that the Church became forced to accept this as fact and do away with its false belief. The idea that Christians believed in a flat Earth has been taught in school textbooks, short films, and is believed by many even today.

Source: http://strangenotions.com/did-the-church-teach-the-earth-was-flat/

Let us thank science as they bring us new insights on the details of what the Bible has always said. *Daniel 12:4 (KJV):* "But thou, O Daniel, shut up the words, and seal the book, even to the time of the end: many shall run to and fro, <u>and knowledge shall be increased</u>."

5. Christ and Water

JESUS AND WATER
Audience: Adult | Youth | Individuals | Small Group Leaders | Church Leaders
Format: Web
Author: Jenny Phillips

Images of water pervade the Gospel stories, symbolizing chaos, rebirth, and new life. Jesus was baptized in water, walked on water, and turned water into wine. These and other narratives are grounded in the stories and experiences of the ancient Israelites, who used ideas about water to better understand their God. To consider Jesus' relationship with water, we must first consider water imagery in the Hebrew Scriptures.

> *In the beginning God created the heavens and the earth. The earth was barren, with no form of life; it was under a roaring ocean covered with darkness. But the Spirit of God was moving over the water* (Genesis 1:1-2, CEV).

In this story, God gives birth to creation by bringing shape and order to watery chaos. This is an important starting point in the story of the ancient Israelites. Water held diverse connotations: from the source of life, to a place of danger, to a means of cleansing and renewal. In the first creation story, God created space between the waters where the earth could flourish. In the Genesis 2 creation story, God began with barren land, and used water as a source of life and renewal for the rest of creation (Genesis 2:6-7). In both instances, whether by separating water from the land or by turning the chaos of water into a source of vitality, God demonstrates divine authority.

Later, in the Genesis flood account, God commands the waters to fill the earth from above and below, sweeping away evil, and temporarily restoring the earth to its pre-creation state. As the waters receded, God's relationship with creation was reborn through a covenant never to destroy the earth by flood again (Genesis 9:11).

Chaos and order. Death and rebirth. These themes from the Hebrew Scriptures are also primary in the Gospels. Jesus began his ministry by stepping into the Jordan River (Matthew 3:13-17). Despite the protests of John the Baptist, Jesus was baptized. When he came up from the water, the heavens opened and the spirit of God descended upon him "like a dove." One might well read this event as a mark of God's entrance into the human experience. Jesus, through his experience in the water, is both diving into the chaos of humanity and demonstrating the cleansing of the soul that comes through God's grace. Whereas in Genesis 1:9-10 God brought order to water, separating the seas from the land, here in Matthew Jesus insists on entering into the waters himself.

We see Jesus' humanity in his self-immersion into the waters of baptism. We also see his divinity as he shows that he too can control water. Jesus quieted a chaotic storm (Mark 4:35-41), walked on water (Matthew 14:22-33), and turned water into wine (John 2:1-11) Taking the water (a reminder of God's first covenant with the creation) and turning it into wine (a symbol of the blood of the new covenant), Jesus said, "If you are thirsty come to me and drink! Have faith in me, and you will have life-giving water flowing from deep inside you, just as the Scriptures say." (John 7:37-39, CEV).

Source: http://bibleresources.americanbible.org/resource/jesus-and-water

6. God made us out of himself. I propose (speculate) that man is therefore part of God both **physically** and **spiritually**. We are just a breath from God, the tiniest part of His Spirit is what we are. ==The breath of God is His Spirit in us, and it is ice.==

> *==The breath of God produces ice,==*
> *==and the broad waters become frozen. (Job 37:10, NIV)==*
> *==But it is the spirit in a man,==*
> *==the breath of the Almighty, that gives him understanding. (Job 32:8, NIV)==*

> *For this is what the high and lofty One says—*
> *he who lives forever, whose name is holy:*
> *"I live in a high and holy place,*
> *but also with him who is contrite and lowly in spirit,*
> *to revive the spirit of the lowly*
> *and to revive the heart of the contrite.*
> *I will not accuse forever,*
> *nor will I always be angry,*
> *==for then the spirit of man would grow faint before me—==*
> *==the breath of man that I have created.==*
> *(Isaiah 57:15–16, NIV)*

And from the Hebrew, showing we come from this breath of the Spirit that is God.

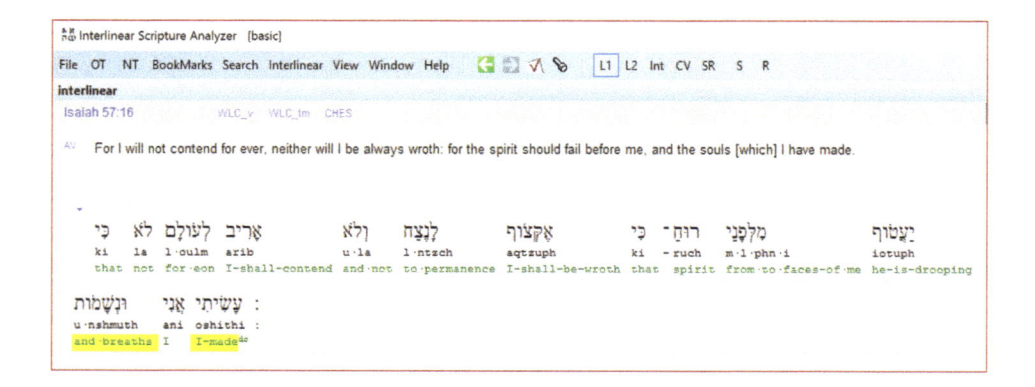

Genesis 2:7 (NIV): "The LORD God formed the man from the dust of the ground **and breathed into his nostrils the breath of life,** *and the man became a living being."* And where did the dust of the ground come from? Christ made it, so what did he use?

> *In the beginning was the Word, and the Word was with God, and the Word was God. He was with God in the beginning.*
> **Through him all things were made; without him nothing was made that has been made. In him was life, and that life was the light of men.** *The light shines in the darkness, but the darkness has not understood it. (John 1:1–5, NIV)*

I believe science can tell us a lot about that. From particle physics, the standard model is a summary of the microworld of dirt. All those subatomic particles are surely there and omnipresent throughout the universe. Just as God is, because I think they are God. I think that matter is like the energy of His Spirit, transformed (like a Higgs boson field would) into mass (dirt). It is my thought that God is both the physical (dirt) and the spirit (just a breath worth) that made us. I propose (speculate) that man is therefore part of God both **physically** and **spiritually**. I believe science is correct with their observations and has seen perhaps the most minute part of physical creation. But they can't see a breath worth of His Spirit. (How many breaths do you take in a lifetime?)

We are like Him, although mostly physically first (first Adam), but eventually also fully spirit (in oneness with the last Adam, Christ).

> *Then God said, "****Let us make man in our image, in our likeness****, and let them rule over the fish of the sea and the birds of the air, over the livestock, over all the earth, and over all the creatures that move along the ground."*
> *So God created man in his own image,*
> *in the image of God he created him;*
> *male and female he created them. (Genesis 1:26–27, NIV)*

Physically first, then spiritual.

> *But Christ has indeed been raised from the dead, the firstfruits of those who have fallen asleep. For since death came through a man, the resurrection of the dead comes also through a man. **For as in Adam all die, so in Christ all will be made alive**. But each in his own turn: Christ, the firstfruits; then, when he comes, those who belong to him. Then the end will come, when he hands over the kingdom to God the Father after he has destroyed all dominion, authority and power. For he must reign until he has put all his enemies under his feet. The last enemy to be destroyed is death. For he "has put everything under his feet." Now when it says that "everything" has been put under him, it is clear that this does not include God himself, who put everything under Christ. When he has done this, then the Son himself will be made subject to him who put everything under him, so that God may be all in all. (1 Corinthians 15:20–28, NIV)*

The Resurrection Body
> *But someone may ask, "How are the dead raised? **With what kind of body will they come**?" How foolish! What you sow does not come to life unless it dies. When you sow, you do not plant the body that will be, but just a seed, perhaps of wheat or of something else. But God gives it a body as he has determined, and to each kind of seed he gives its own body. All flesh is not the same: Men have one kind of flesh, animals have another, birds another and fish another. There are also heavenly bodies and there are earthly bodies; but the splendor of the heavenly bodies is one kind, and the splendor of the earthly bodies is another. The sun has one kind of splendor, the moon another and the stars another; and star differs from star in splendor.*

So will it be with the resurrection of the dead. **The body that is sown is perishable, it is raised imperishable; it is sown in dishonor, it is raised in glory;** *it is sown in weakness, it is raised in power;* ==**it is sown a natural body, it is raised a spiritual body.**==

If there is a natural body, there is also a spiritual body. So it is written: "The first man Adam became a living being"; the last Adam, a life-giving spirit. The spiritual did not come first, but the natural, and after that the spiritual. The first man was of the dust of the earth, the second man from heaven. As was the earthly man, so are those who are of the earth; and as is the man from heaven, so also are those who are of heaven. And just as we have borne the likeness of the earthly man, so shall we bear the likeness of the man from heaven.

I declare to you, brothers, ==**that flesh and blood cannot inherit the kingdom of God**==**, nor does the perishable inherit the imperishable. Listen, I tell you a mystery: We will not all sleep,** ==**but we will all be changed**==**—in a flash, in the twinkling of an eye, at the last trumpet. For the trumpet will sound, the dead will be raised imperishable,** ==**and we will be changed.**== *For the perishable must clothe itself with the imperishable, and the mortal* ==**with immortality**==*. When the perishable has been clothed with the imperishable, and the mortal with immortality, then the saying that is written will come true: "Death has been swallowed up in victory." (1 Corinthians 15:35–54, NIV)*

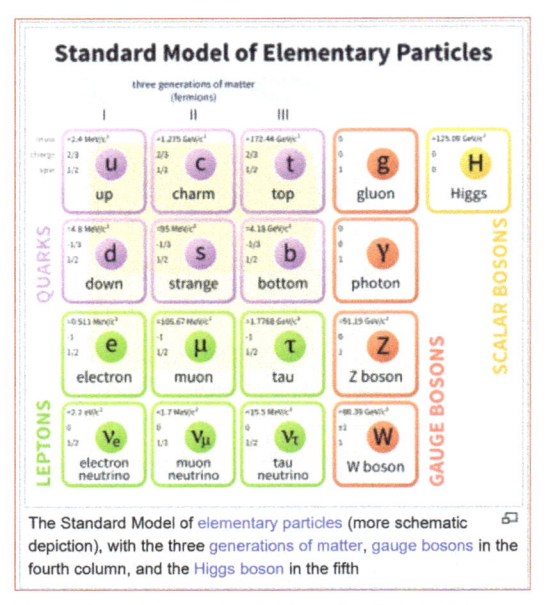

Source: https://en.wikipedia.org/wiki/Standard_Model

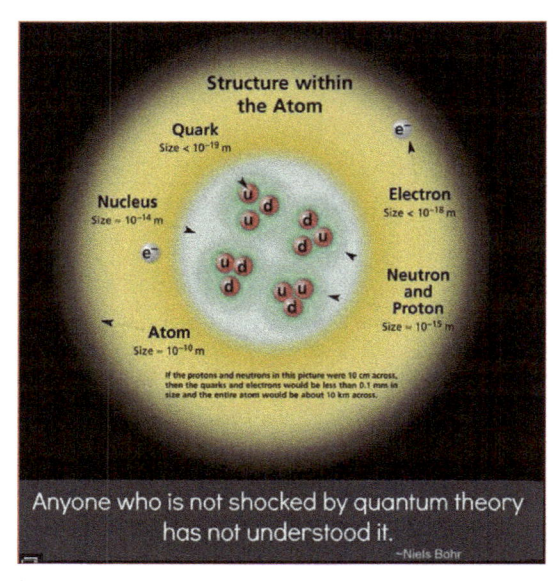

https://www.bing.com/images/search?view=detailV2&ccid=FCVRVLX
i&id=957042F6922477B149C2A7C03BF58D5074CB4D49&thid=
OIP.FCVRVLXiM1v1l5bdx6iUpgEdEs&q=standard+model+physics&
simid=608027470768834232&selectedIndex=47&ajaxhist=0

7. Time not absolute

Einstein proved that when you move fast, time slows down, which was not understood by man until the twentieth century. Yet it makes what God says literally possible, as well as figuratively true.

If you were to hypothetically rocket away from a friend on earth very quickly, depending on your speed and path, it could take a thousand years until your return (according to the one left behind on earth), but according to you, it might only be a day later! You would be astonished that the person you left behind died centuries ago. For time is relative and not absolute, as previously thought through most of recorded history.

Now please consider, God would have been with you both for the duration. God therefore saw your friend age, have grandchildren, and finally die, as He watched you hurtle through space for only a day, perhaps not even sleeping. Right now, that is just a thought experiment because until your change from physical to spirit, you couldn't travel that fast. But God could make that trip, and a thousand years would be as a day! Come to think of it, if God wanted to, I think God could temporarily change you to spirit and bring you along!

> *__I was in the Spirit on the Lord's day__, and I heard a loud voice behind me like a trumpet saying, "Write on a scroll what you see and send it to the seven churches: Ephesus, Smyrna, Pergamum, Thyatira, Sardis, Philadelphia, and Laodicea." (Revelation 1:10–11, HCSB)*

In the Greek, we see John telling us that he "became in the spirit IN the Lord's Day." Note that it says "in" and not "on." The New Testament church was started on the first day of the week (a Sunday, special High Day of Pentecost AD 31), but I think this is talking about John's presence in a more general greater duration—time of God's punishment on His return to earth, the day of the Lord's vengeance, and not a day of the week in my mind. Besides, God's weekly day is the Sabbath (Saturday), so that would be the time set out as special through the week, or the Lord's time for special services or the weekly day of the Lord.

And what difference would the day of the week make? The writings are about a time when God makes things happen and is no longer silent. The day of the Lord is talked about a lot in other prophecies throughout the Bible. And he asked John to write the events of this special time in history for a permanent prerecord.

Not just saying, by the way it was Saturday (or Sunday, the first day of the new week).

> ***The Day of the LORD***
> *Woe because of that day!*
> *For the Day of the LORD is near*
> *and will come as devastation from the Almighty.*
> *(Joel 1:15, HCSB)*
>
> *Look,* ***the day of the LORD*** *is coming—*
> *cruel, with rage and burning anger—*
> *to make the earth a desolation*
> *and to destroy the sinners on it.*
> *Indeed, the stars of the sky and its constellations*
> *will not give their light.*
> *The sun will be dark when it rises,*
> *and the moon will not shine.*
> *I will bring disaster on the world,*
> *and their own iniquity, on the wicked.*
> *I will put an end to the pride of the arrogant*
> *and humiliate the insolence of tyrants.*
> *(Isaiah 13:9–11, HCSB)*

There are about eighty-six quotes about the day of the Lord. This is not just some off-the-cuff discussion about a day of the week.

SPACE THE TRUE FRONTIER!

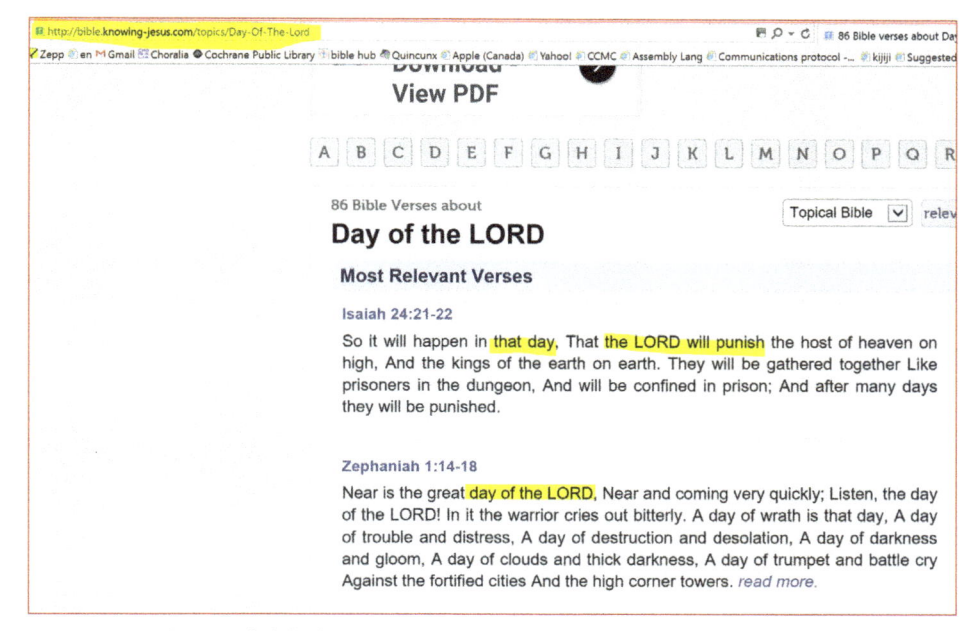

> **86 Bible Verses about**
> **Day of the LORD**
>
> **Most Relevant Verses**
>
> **Isaiah 24:21-22**
> So it will happen in that day, That the LORD will punish the host of heaven on high, And the kings of the earth on earth. They will be gathered together Like prisoners in the dungeon, And will be confined in prison; And after many days they will be punished.
>
> **Zephaniah 1:14-18**
> Near is the great day of the LORD, Near and coming very quickly; Listen, the day of the LORD! In it the warrior cries out bitterly. A day of wrath is that day, A day of trouble and distress, A day of destruction and desolation, A day of darkness and gloom, A day of clouds and thick darkness, A day of trumpet and battle cry Against the fortified cities And the high corner towers. *read more.*

Source: http://bible.knowing-jesus.com/topics/Day-Of-The-Lord

    I think John did some fast movement through time with God and witnessed future drastic events of the Lord's Day in the future as they occur and was not just seeing a holodeck vision. This would be possible with a God that lives outside of time and anyone He brings into that reality (even temporarily).

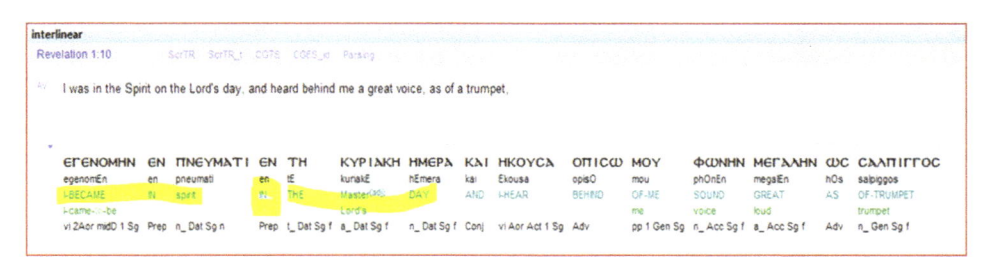

> **interlinear**
> **Revelation 1:10**
>
> I was in the Spirit on the Lord's day, and heard behind me a great voice, as of a trumpet.
>
> ΕΓΕΝΟΜΗΝ ΕΝ ΠΝΕΥΜΑΤΙ ΕΝ ΤΗ ΚΥΡΙΑΚΗ ΗΜΕΡΑ ΚΑΙ ΗΚΟΥΣΑ ΟΠΙΣΩ ΜΟΥ ΦΩΝΗΝ ΜΕΓΑΛΗΝ ΩΣ ΣΑΛΠΙΓΓΟΣ
> egenomEn en pneumati en tE kuriakE hEmera kai Ekousa opisO mou phOnEn megalEn hOs salpiggos
> I-BECAME IN spirit IN THE Master(s) DAY AND I-HEAR BEHIND OF-ME SOUND GREAT AS OF-TRUMPET
> I-came---be Lord's me voice loud trumpet
> vi 2Aor midD 1 Sg Prep n_ Dat Sg n Prep t_ Dat Sg f a_ Dat Sg f n_ Dat Sg f Conj vi Aor Act 1 Sg Adv pp 1 Gen Sg n_ Acc Sg f a_ Acc Sg f Adv n_ Gen Sg f

245

8. God is everywhen? *"And this one thing let not be unobserved by you, beloved, that one day with the Lord is as a thousand years, **and a thousand years as one day**" (2 Peter 3:8, YLT).* I find it more than interesting that the theory of relativity, has shown that moving away from the earth at high speed slows down time. Is that not the realm of God? Could this again be science confirming what is scriptural? Time is not absolute, and it would certainly not limit the one who created it.

If God is not bound by time that he created, would He not therefore exist at both the beginning and the end. I think of it like God is at the hub of a wheel, connected by the spokes to every part of the wheel rim. We travel out on the rim requiring time to move from place to place, but God at the hub, is at all points at once (everywhere and everywhen).

> *Remember the former things, those of long ago;*
> *I am God, and there is no other;*
> *I am God, and there is none like me.*
> ***I make known the end from the beginning,***
> *from ancient times, what is still to come.*
> *I say: My purpose will stand,*
> *and I will do all that I please. (Isaiah 46:9–10, NIV)*

> *Which God will bring about in his own time—God, the blessed and only Ruler, the King of kings and Lord of lords, **who alone is immortal** and who lives in unapproachable light, whom no one has seen or can see. To him be honor and might forever. Amen. (1 Timothy 6:15–16, NIV)*

> *Lord, you have been our dwelling place*
> *throughout all generations.*
> *Before the mountains were born*
> *or you brought forth the earth and the world,*
> ***from everlasting to everlasting you are God.***
> *You turn men back to dust,*
> *saying, "Return to dust, O sons of men."*

For a thousand years in your sight
are like a day that has just gone by,
or like a watch in the night. *(Psalms 90:1–4, NIV)*

For this is what the high and lofty One says—
he who lives forever, whose name is holy:
"I live in a high and holy place,
but also with him who is contrite and lowly in spirit,
to revive the spirit of the lowly
and to revive the heart of the contrite.
(Isaiah 57:15, NIV)

Now to the King eternal, immortal, invisible, the only God, be honor and glory for ever and ever. Amen.
(1 Timothy 1:17, NIV)

9. The Universes Decay: Entropy

Below is a discussion from the Internet that I include because it speaks of the restoration and the new heavens that are coming. I, however, disagree that the decay comes from man. Science shows that it is part of creation's natural winding down, or increasing entropy. I very much agree, that God has a permanent fix that will ensure eternity. (http://www.biblearchaeology.org/post/2012/12/26/The-Crumbling-of-Creation-the-Cause-of-the-Earths-Decay-and-Gods-Glorious-Cure.aspx#Article)

God's Glorious Cure: The New Heavens and New Earth

The glorious redemption both of mankind and all of creation is the high point of Paul's discussion in Rom 8:18-25. Both believers in Christ and all of creation are currently groaning under sin and God's curse. But there awaits a glorious redemption for both man and creation.[23]

Many of the OT prophets contain this theme of creation's glorious restoration. Jeremiah 33:12-16 speaks of the desolate place of Jerusalem becoming inhabited once more, once the Messiah, the "Branch of righteousness," is ruling righteously on the throne and Judah is saved. In Ezek 34:23-31, once the Lord and His servant David are on the throne, then there will be "showers of blessing" on the land, and the trees and the earth will again be prolific with their produce. In Ezek 36, the desolate mountains (compare Ezek 6:1-10) will now shoot forth their branches and yield their fruit to Israel (vv. 8-9), and the waste places will become like the garden of Eden (vv. 33-36). In Hos 2:21-23, instead of the Lord withholding grain, new wine, and oil, the land will overflow with them (the same thought is expressed in Joel 2:19, 22-24). Amos 9:13-15 speaks of the fertility of the land, with abundant wine and fruit. And Zech 8:12 speaks of the prosperity of the land once the Lord restores Israel and dwells in her midst (Zech 8:3-8).

But no prophet speaks of the restoration of creation as eloquently or as frequently as the prophet Isaiah. In Isa 11, once the righteous Ruler is on the throne, the entire earth will be filled with the knowledge of the Lord (v. 9). The animals will no longer need to be feared, but will dwell peaceably with little children (vv. 6-9).[24]

Conclusion

Yes, the earth is decaying. And yes, the cause is man's sin. And it is certainly true that man may contribute to the decay of God's creation by not being good stewards of creation as God intended –whether by pollution of the air, deforestation, overutilization of some resources, contributing to species extinction, and so forth.[30] But that is not the emphasis of the Scriptures, from Genesis to Revelation. The emphasis of these passage is that man has disobeyed God, and therefore the creation also has come under judgment. The ultimate answer is not to eliminate mankind from the planet, or to enact hundreds of environmental regulations (though some may be needed). The ultimate answer is for man to be restored to God.[31] And God promises to do exactly that. The same biblical passages that speak of God's judgment upon man and creation also speak of a glorious restoration. Christ will return to judge the wicked and redeem the righteous, and there will be a new heavens and a new earth. At that time creation will be completely restored. Man once again will rule (with Christ as the head), and both man and creation will be in peace and harmony at last, all according to the grand purpose of God. May that time come quickly! Maranatha!

Todd S. Beall, Ph.D. is Professor of Old Testament Literature & Exegesis at Capital Bible Seminary, Langham, MD, since 1977. He served as Assistant Academic Dean and Registrar, from 1978–94. He was a Translator/Editor of the Holman Christian Standard Bible, 1998–2002. Dr. Beall formerly served on the ABR Board of Directors for 15 years.

The universe is becoming disordered and, as is, will not exist eternally. Science has shown us why we need a new creation because this one is like a clock that is winding down. It will eventually reach absolute zero and does not last forever. *Matthew 24:35 (KJV):* "**Heaven and earth shall pass away**, *but my words shall not pass away.*" God will fix that, and there will be no such decay in the future.

> *And I saw a new heaven and a new earth:* **для the first heaven and the first earth were passed away**; *and there was no more sea. (Revelation 21:1, KJV)*
>
> ***The Day of the Lord***
> *Dear friends, this is now my second letter to you. I have written both of them as reminders to stimulate you to wholesome thinking. I want you to recall the words spoken in the past by the holy prophets and the command given by our Lord and Savior through your apostles.*
> *First of all, you must understand that in the last days scoffers will come, scoffing and following their own evil desires. They will say,* "**Where is this 'coming' he promised**? *Ever since our fathers died, everything goes on as it has since the beginning of creation."* **But they deliberately forget** *that* **long ago by God's word the heavens existed and the earth was formed out of water and by water.** *By these waters also the world of that time was deluged and destroyed.* **By the same word the present heavens and earth are reserved for fire, being kept for the day of judgment and destruction of ungodly men.**
> *But do not forget this one thing, dear friends:* **With the Lord a day is like a thousand years, and a thousand years are like a day.** *The Lord is not slow in keeping his promise, as some understand slowness. He is patient with you, not wanting anyone to perish, but everyone to come to repentance.*

But the day of the Lord will come like a thief. ***The heavens will disappear with a roar; the elements will be destroyed by fire****, and the earth and everything in it will be laid bare.*

Since everything will be destroyed in this way, what kind of people ought you to be? You ought to live holy and godly lives [as you look forward to the day of God and speed its coming. That day will bring about the destruction of the heavens by fire, ***and the elements will melt in the heat. But in keeping with his promise we are looking forward to a new heaven and a new earth, the home of righteousness.***

So then, dear friends, since you are looking forward to this, make every effort to be found spotless, blameless and at peace with him. Bear in mind that our Lord's patience means salvation, just as our dear brother Paul also wrote you with the wisdom that God gave him. He writes the same way in all his letters, speaking in them of these matters. His letters contain some things that are hard to understand, which ignorant and unstable people distort, as they do the other Scriptures, to their own destruction.

Therefore, dear friends, since you already know this, be on your guard so that you may not be carried away by the error of lawless men and fall from your secure position. ***But grow in the grace and knowledge of our Lord and Savior Jesus Christ. To him be glory both now and forever! Amen.*** *(2 Peter 3:1–18, NIV)*

SPACE THE TRUE FRONTIER!

The laws of thermodynamics define fundamental physical quantities (temperature, energy, and entropy) that characterize thermodynamic systems.

LEARNING OBJECTIVE

- Discuss the three laws of thermodynamics.

KEY POINTS

- The first law, also known as Law of Conservation of **Energy**, states that energy cannot be created or destroyed in an **isolated system**.
- The second law of **thermodynamics** states that the **entropy** of any isolated system always increases.
- The third law of thermodynamics states that the entropy of a system approaches a constant value as the **temperature** approaches **absolute zero**.

TERMS

- absolute zero

  The lowest temperature that is theoretically possible.

- entropy

  A thermodynamic property that is the measure of a system's thermal energy per unit of temperature that is unavailable for doing useful work.

Source: https://www.boundless.com/chemistry/textbooks/boundless-chemistry-textbook/thermodynamics-17/the-laws-of-thermodynamics-123/the-three-laws-of-thermodynamics-496-3601/

en·tro·py
[ˈentrəpē]

NOUN

1. a thermodynamic quantity representing the unavailability of a system's thermal energy for conversion into mechanical work, often interpreted as the degree of disorder or randomness in the system.
2. <mark>lack of order or predictability; gradual decline into disorder:</mark>
"a marketplace where entropy reigns supreme"
synonyms: deterioration · degeneration · crumbling · decline · [more]
3. (in information theory) a logarithmic measure of the rate of transfer of information in a particular message or language.

Source: https://www.bing.com/search?q=entropy&src=IE-SearchBox&FORM=IESR02&pc=EUPP_

10. Ancient night skies (what man could guess was there).

Man, on his own, and without the benefits of scientific understanding or God-ordained explanation would make some understandably wild guesses. Yet the Bible says nothing that contradicts what science observes. God still sits above the circle (or sphere) of the earth, just like He said in the book of Job thousands of years ago.

> *Isn't God as high as the heavens? And look at the highest stars —how lofty they are! Yet you say: "What does God know? Can He judge through thick darkness? Clouds veil Him so that He cannot see, as He walks on the circle of the sky." (Job 22:12–14, HCSB)*

SPACE THE TRUE FRONTIER!

Grizzly Bear *Cygnus   Milky Way*
Shoshone Tribe
Wyoming, Southern Idaho

A grizzly bear (Cygnus) climbed up a tall mountain to go hunting in the sky. As he climbed, snow and ice clung to the fur of his feet and legs. Crossing the sky the ice crystals trailed behind him forming the Milky Way.

Elk Skin *Cassiopeia*
Yakima Tribe
Central Washington

A Hunter killed a great elk and stretched the skin to dry by driving wooden stakes through it. Afterwards he threw the skin into the sky (Cassiopeia) where the light above shines through the stake holes forming stars.

Coyote's Eyeball *Arcturus*
Lummi Tribe
Pacific Northwest Coast

The Coyote liked to take out his eyeballs and juggle them to impress the girls. One day as he was juggling them he threw one so high it stuck in the sky (Arcturus).

Myths about the Sky, Constellations, and Stars

👍 Like 74

Since long ago, people around the world have associated the heavens, the stars, and the patterns they make in the sky with their gods and goddesses. Links from this page will take you to descriptions of the role of selected stars, star patterns, and related gods and goddesses in various cultures.

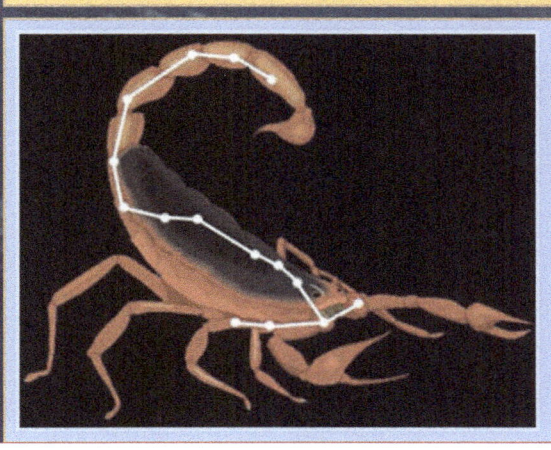

This is the Greek constellation Scorpius, the scorpion. It contains a bright star, Antares, that is often referred to as "the heart of the scorpion". To the ancient Greeks, Scorpius was related to the death of the hunter Orion.

Source: http://www.wwu.edu/depts/skywise/legends.html

11. The second death

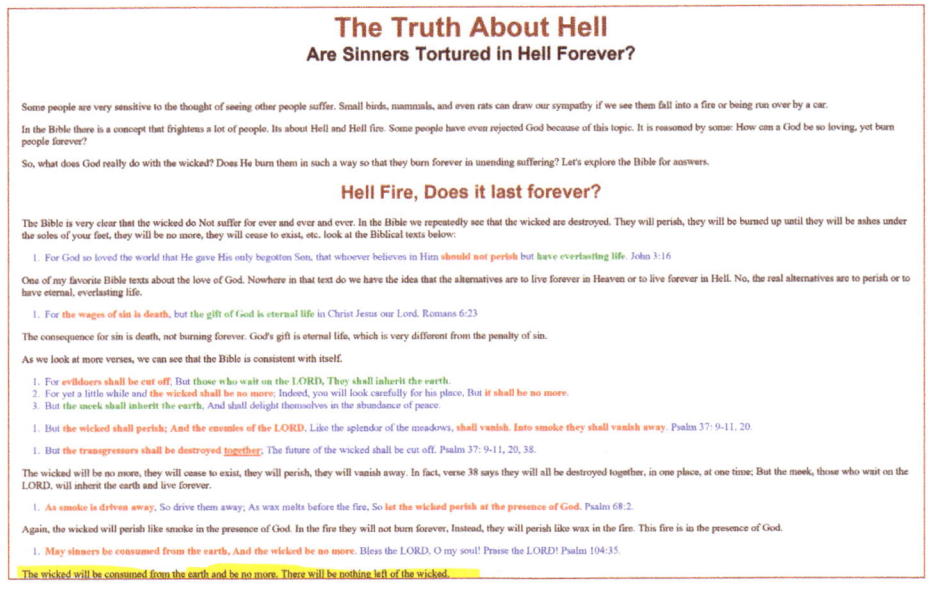

Source: http://www.creation-science-prophecy.com/hell.htm

12. Circle of the earth, may mean more of a **vault** than a sphere

*Thick clouds are a covering to him, so that he seeth not;
And he walketh **on the vault** of heaven. (Job 22:14, ASV)*

Source: *https://www.bibleandscience.com/bible/books/genesis/genesis1_circleearth.htm*

SPACE THE TRUE FRONTIER!

13. The New creation, and Israel of God today Galations 6:16

Jesus never said He was King of the Jews. That was Pilate. Christ said He was a King, true, but He is King of all Israel, not just Judah!

John 18:33 (HCSB) [33] Then Pilate went back into the headquarters, summoned Jesus, and said to Him, "Are You the King of the Jews?"

[34] Jesus answered, "Are you asking this on your own, or have others told you about Me?"

[35] "I'm not a Jew, am I?" Pilate replied. "Your own nation and the chief priests handed You over to me. What have You done?"

[36] "My kingdom is not of this world," said Jesus. "If My kingdom were of this world, My servants would fight, so that I wouldn't be handed over to the Jews. As it is, My kingdom does not have its origin here."

[37] "You are a king then?" Pilate asked.

"You say that I'm a king," Jesus replied. "I was born for this, and I have come into the world for this: to testify to the truth. Everyone who is of the truth listens to My voice."

Where it began:
Genesis 32:24 (HCSB)
Jacob Wrestles with God
[24] Jacob was left alone, and a man wrestled with him until daybreak. [25] When the man saw that He could not defeat him, He struck Jacob's hip socket as they wrestled and dislocated his hip. [26] Then He said to Jacob, "Let Me go, for it is daybreak."

But Jacob said, "I will not let You go unless You bless me."

[27] "What is your name?" the man asked.

"Jacob," he replied.

[28] "Your name will no longer be Jacob," He said. "It will be Israel because you have struggled with God and with men and have prevailed."
[29] Then Jacob asked Him, "Please tell me Your name."
But He answered, "Why do you ask My name?" And He blessed him there.
[30] Jacob then named the place Peniel, "For I have seen God face to face," he said, "and I have been delivered." [31] The sun shone on him as he passed by Penuel — limping because of his hip. [32] That is why, to this day, the Israelites don't eat the thigh muscle that is at the hip socket: because He struck Jacob's hip socket at the thigh muscle.
And Today: (We overcome/prevail as the bride of Christ)
Israel, overcomers through Christ in us. We have wrestled between God in us, and the man we are. And Christ prevails.
Romans 8:8 (HCSB)
[8] Those who are in the flesh cannot please God. [9] You, however, are not in the flesh, but in the Spirit, since the Spirit of God lives in you. But if anyone does not have the Spirit of Christ, he does not belong to Him. [10] Now if Christ is in you, the body is dead because of sin, but the Spirit is life because of righteousness. [11] And if the Spirit of Him who raised Jesus from the dead lives in you, then He who raised Christ from the dead will also bring your mortal bodies to life through His Spirit who lives in you. More on the eternal Kingdom of Israel:
https://www.ucg.org/the-good-news/the-kingdom-of-god-the-heart-of-christs-message

SPACE THE TRUE FRONTIER!

Christ knew He was King

The bold prophecies of the Old Testament, combined with the Gospels and apostolic writings of the New, show that Christians ought to understand Christ both as Savior and as returning King (Daniel 7:13-14; Revelation 11:15; Acts 1:1 11).

It was Jesus' deliberate association of Himself with the prophecy in Daniel that inflamed the chief priests and settled the resolve of the Sanhedrin that Jesus must die (Mark 14:53, 61-65).

Jesus did not want His followers to expect an immediate kingdom (Luke 19:11-12), but He did purposefully cultivate their expectation for a kingdom in the future with Himself as its Monarch. The act of cutting and scattering tree branches that marked His entrance into Jerusalem shortly before His death was done as a tribute to the Messiah King in fulfillment of Zechariah's prophecy. This significance was not lost on the Pharisees (Matthew 21:1 11; Luke 19:28-40).

This teaching cost Jesus His life. When He was soon to become the Messiah Savior through His death and resurrection, He would not deny being Messiah King (John 18:33-37). When Pilate questioned Him about His kingship, Christ replied, "For this cause I was born" (verse 37).

This became the focus of the soldiers' ridicule and torture (John 19:1 3). It also formed the closing argument of Christ's accusers, who used this seditious charge to force Pilate into issuing the order to have Jesus put to death (John 19:12 16). The derisive comments made to Jesus as He was crucified further confirm He was killed because He claimed to be the Messiah King (Mark 15:31 32). Jesus' kingship was clearly marked on the sign above Him as He died (John 19:19).

Continuing Kingdom message

Both Jesus and the writers of the New Testament had countless opportunities to sever links to expectations of a coming king and kingdom spoken of by the Old Testament prophets. But, rather than cut those links, Christ and His disciples deliberately built upon them.

Gods plan to increase His Family, by adopting sons, continues. This Kingdom for Israel is closer. The times of refreshing will come. To Israel:
Galatians 3:29 (NIV)
[29] If you belong to Christ, then you are Abraham's seed, and heirs according to the promise.

Acts 3:11 (NIV)
Peter Speaks to the Onlookers
[11] While the beggar held on to Peter and John, all the people were astonished and came running to them in the place called Solomon's Colonnade. [12] When Peter saw this, he said to them: "Men of Israel, why does this surprise you? Why do you stare at us as if by our own power or godliness we had made this man walk? [13] The God of Abraham, Isaac and Jacob, the God of our fathers, has glorified his servant Jesus. You handed him over to be killed, and you disowned him before Pilate, though he had decided to let him go. [14] You disowned the Holy and Righteous One and asked that a murderer be released to you. [15] You killed the author of life, but God raised him from the dead. We are witnesses of this. [16] By faith in the name of Jesus, this man whom you see and know was made strong. It is Jesus' name and the faith that comes through him that has given this complete healing to him, as you can all see.
[17] "Now, brothers, I know that you acted in ignorance, as did your leaders. [18] But this is how God fulfilled what he had foretold through all the prophets, saying that his Christ would suffer. [19] Repent, then, and turn to God, so that your sins may be wiped out, that times of refreshing may come from the Lord, [20] and that he may send the Christ, who has been appointed for you—even Jesus. [21] He must remain in heaven until the time comes for God to restore everything, as he promised long ago through his holy prophets. [22] For Moses said, 'The Lord your God will raise up for you a prophet like me from among your own people; you must listen to everything he tells you. [23] Anyone who does not listen to him will be completely cut off from among his people.'

[24] "Indeed, all the prophets from Samuel on, as many as have spoken, have foretold these days. [25] And you are heirs of the prophets and of the covenant God made with your fathers. He said to Abraham, 'Through your offspring all peoples on earth will be blessed.' [26] When God raised up his servant, he sent him first to you to bless you by turning each of you from your wicked ways."

Christ was King of the Jews, and King of all Israel. Judah (Jews) were only one of the 12 tribes. Christ will rule over 12 tribes of Israel, yet, with us: Luke 22:29 (NIV)

[29] And I confer on you a kingdom, just as my Father conferred one on me, [30] so that you may eat and drink at my table in my kingdom and sit on thrones, judging the twelve tribes of Israel.

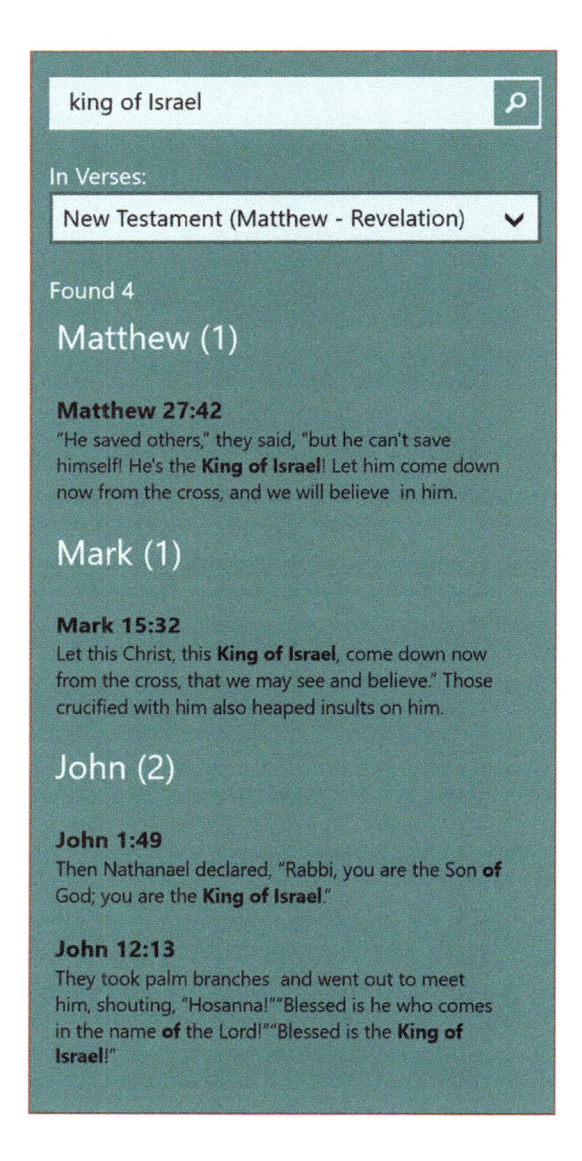

> **king of Israel**
>
> In Verses:
> New Testament (Matthew - Revelation)
>
> Found 4
>
> ## Matthew (1)
>
> **Matthew 27:42**
> "He saved others," they said, "but he can't save himself! He's the **King of Israel**! Let him come down now from the cross, and we will believe in him.
>
> ## Mark (1)
>
> **Mark 15:32**
> Let this Christ, this **King of Israel**, come down now from the cross, that we may see and believe." Those crucified with him also heaped insults on him.
>
> ## John (2)
>
> **John 1:49**
> Then Nathanael declared, "Rabbi, you are the Son **of** God; you are the **King of Israel**."
>
> **John 12:13**
> They took palm branches and went out to meet him, shouting, "Hosanna!""Blessed is he who comes in the name **of** the Lord!""Blessed is the **King of Israel!**"

Jesus is King of Israel, the people of God (transcending the covenants).

The Birth of Jesus
In those days Caesar Augustus issued a decree that a census should be taken of the entire Roman world. (This was the first census that took place while Quirinius was

governor of Syria.) And everyone went to his own town to register.

So Joseph also went up from the town of Nazareth in Galilee to Judea, to Bethlehem the town of David, because **he belonged to the house and line of David***. He went there to register with Mary, who was pledged to be married to him and was expecting a child. While they were there, the time came for the baby to be born, and she gave birth to her firstborn, a son.* **She wrapped him in cloths and placed him in a manger,** *because there was no room for them in the inn.*

The Shepherds and the Angels

And there were shepherds living out in the fields nearby, keeping watch over their flocks at night. An angel of the Lord appeared to them, and the glory of the Lord shone around them, and they were terrified. But the angel said to them, "Do not be afraid. I bring you good news of great joy that will be for all the people. **Today in the town of David a Savior has been born to you; he is Christ the Lord.** *This will be a sign to you: You will find a baby wrapped in cloths and lying in a manger."*

Suddenly a great company of the heavenly host appeared with the angel, praising God and saying,

"Glory to God in the highest,

and on earth peace to men on whom his favor rests."

When the angels had left them and gone into heaven, the shepherds said to one another, "Let's go to Bethlehem and see this thing that has happened, which the Lord has told us about."

So they hurried off and found Mary and Joseph, and the baby, who was lying in the manger. When they had seen him, they spread the word concerning what had been told them about this child, and all who heard it were amazed at what the shepherds said to them. But Mary treasured up all these things and pondered them in her heart. The shepherds returned, glorifying and praising

God for all the things they had heard and seen, which were just as they had been told.
Jesus Presented in the Temple
On the eighth day, when it was time to circumcise him, he was named Jesus, the name the angel had given him before he had been conceived.
When the time of their purification according to the Law of Moses had been completed, *Joseph and Mary took him to Jerusalem to present him to the Lord* **(as it is written in the Law of the Lord, "Every firstborn male is to be consecrated to the Lord"),** *and to offer a sacrifice in keeping with what is said in the Law of the Lord: "a pair of doves or two young pigeons."*
Now there was a man in Jerusalem called Simeon, who was righteous and devout. He was waiting for the consolation of Israel, and the Holy Spirit was upon him. **It had been revealed to him by the Holy Spirit that he would not die before he had seen the Lord's Christ.** *Moved by the Spirit, he went into the temple courts. When the parents brought in the child Jesus to do for him what the custom of the Law required, [28] Simeon took him in his arms and praised God, saying:*
"Sovereign Lord, as you have promised,
you now dismiss your servant in peace.
For my eyes have seen your salvation,
which you have prepared in the sight of all people,
a light for revelation to the Gentiles
and for glory to your people Israel."
The child's father and mother marveled at what was said about him. Then Simeon blessed them and said to Mary, his mother: **"This child is destined to cause the falling and rising of many in Israel,** *and to be a sign that will be spoken against, so that the thoughts of many hearts will be revealed. And a sword will pierce your own soul too."*

There was also a prophetess, Anna, the daughter of Phanuel, of the tribe of Asher. She was very old; she had lived with her husband seven years after her marriage, and then was a widow until she was eighty-four. She never left the temple but worshiped night and day, fasting and praying. Coming up to them at that very moment, she gave thanks to God and spoke ==about the child to all who were looking forward to the redemption of Jerusalem. When Joseph and Mary had done everything required by the Law of the Lord, they returned to Galilee to their own town of Nazareth.== *(Luke 2:1–39 (NIV)*

The Old Testament requirements were fulfilled; now the destined future salvation of Israel could begin—a handoff from the Old Testament to the New. Songs of the old sang and honored in the fulfilment of the new song of the Lamb. God's people through history were not Christians until Christ came. But they always were and always will be the rulers with God (israel).

The law of Moses was the schoolmaster until the better and new law of Christ.

Seven Angels With Seven Plagues
I saw in heaven another great and marvelous sign: seven angels with the seven last plagues—last, because with them God's wrath is completed. **And I saw what looked like a sea of glass mixed with fire and, standing beside the sea, those who had been victorious over the beast and his image and over the number of his name. They held harps given them by God** ==and sang the song of Moses the servant of God and the song of the Lamb.== *(Revelation 15:1–3, NIV)*

And in answer to John the Baptist's question, how do you know the Messiah?

In Praise of John the Baptist
When Jesus had finished giving orders to His 12 disciples, He moved on from there to teach and preach in their towns. When John heard in prison what the Messiah was doing, he sent a message by his disciples and asked Him, "Are You the One who is to come, or should we expect someone else?"

Jesus replied to them, <u>**"Go and report to John what you hear and see: the blind see, the lame walk, those with skin diseases are healed, the deaf hear, the dead are raised, and the poor are told the good news.**</u> *And if anyone is not offend(ed) because of Me, he is blessed."*

As these men went away, Jesus began to speak to the crowds about John: "What did you go out into the wilderness to see? A reed swaying in the wind? What then did you go out to see? A man dressed in soft clothes? Look, those who wear soft clothes are in kings' palaces. But what did you go out to see? A prophet? Yes, I tell you, and far more than a prophet. This is the one it is written about:

Look, I am sending My messenger ahead of You; he will prepare Your way before You.

"I assure you: Among those born of women no one greater than John the Baptist has appeared, but the least in the kingdom of heaven is greater than he. From the days of John the Baptist until now, the kingdom of heaven has been suffering violence, and the violent have been seizing it by force. For all the prophets and the Law prophesied until John; if you're willing to accept it, he is the Elijah who is to come. Anyone who has ears should listen! (Matthew 11:1–15, HCSB)

Great crowds came to him, bringing the lame, the blind, the crippled, the mute and many others, and laid them at his feet; and he healed them. **The peo-**

ple were amazed when they saw the mute speaking, the crippled made well, the lame walking and the blind seeing. <mark>*And they praised the God of Israel.*</mark>
(Matthew 15:30–31, NIV)

I propose that this Jesus in front of them, who prophetically healed the blind, was the i am, God of Israel, who became man and walked the earth.

And I say, this Israel is not some historical people. This was written for the eternal Israel of God into our future.

"Great and marvelous are your deeds,
Lord God Almighty.
Just and true are your ways,
King of the ages.
Who will not fear you, O Lord,
and bring glory to your name?
For you alone are holy.
All nations will come
and worship before you,
for your righteous acts have been revealed."
After this I looked and in heaven the temple, that is, the tabernacle of the Testimony, was opened. Out of the temple came the seven angels with the seven plagues. They were dressed in clean, shining linen and wore golden sashes around their chests. Then one of the four living creatures gave to the seven angels seven golden bowls filled with the wrath of God, who lives for ever and ever. And the temple was filled with smoke from the glory of God and from his power, and no one could enter the temple until the seven plagues of the seven angels were completed. (Revelation 15:3–18)

Christ is in us and, therefore, so are the everlasting promises to the seed.

The promises were spoken to Abraham and to his seed. The Scripture does not say "and to seeds," meaning many people, but "and to your seed," meaning one person, who is Christ. What I mean is this: The law, introduced 430 years later, does not set aside the covenant previously established by God and thus do away with the promise. For if the inheritance depends on the law, then it no longer depends on a promise; **but God in his grace gave it to Abraham through a promise.**

What, then, was the purpose of the law? It was added because of transgressions until the Seed to whom the promise referred had come. The law was put into effect through angels by a mediator. A mediator, however, does not represent just one party; but God is one.

Is the law, therefore, opposed to the promises of God? Absolutely not! For if a law had been given that could impart life, then righteousness would certainly have come by the law. But the Scripture declares that the whole world is a prisoner of sin, so that what was promised, being given through faith in Jesus Christ, might be given to those who believe.

Before this faith came, we were held prisoners by the law, locked up until faith should be revealed. **So the law was put in charge to lead us to Christ that we might be justified by faith. Now that faith has come, we are no longer under the supervision of the law.**
Sons of God
You are all sons of God through faith in Christ Jesus, for all of you who were baptized into Christ have clothed yourselves with Christ. **There is neither Jew nor Greek, slave nor free, male nor female, for you are all one in Christ Jesus.** *If you belong to Christ, then you are Abraham's seed, and heirs according to the promise. (Galatians 3:16–29, NIV)*

You are the Israel of the new covenant that is founded on the actions of Christ and not dependent on our actions, in which we proved our failure. But God did not fail His people, ISRAEL, are therein saved. That promise for God's people was and is and always will be.

14. The sign of the Son of Man, note it is Christ speaking.

> Christ, our Passover, probably died just as the lambs were being slaughtered.
> For as Jonah was in the belly of the huge fish three days and three nights, so **the Son of Man will be in the heart of the earth three days and three nights.** (Matthew 12:40, HCSB)

Picture of Passover week (probably AD 31), **Passover on the fourteenth of Nissan**, the **First** and **Last High Days** of Unleavened Bread were the fifteenth and twenty-first respectively. This shows the best way to fit three days and three nights. It has Christ dying on the Passover Wednesday (not Friday) at the same time the regular lambs would be slain by the high priests at the temple that year (see below commentary from F. LaGard Smith in the *Daily Chronological Bible*). Mr. Smith has the timing a little different, but he makes a most excellent point in his commentary.

But I cannot prove my timing shown crudely below. It fits all scriptures except when they are on the road to Emmaus and they say it is the third day (this diagram would have that as the fourth day). God will have to clear this up; it is the best I can figure at this time (September 16, 2017).

Notwithstanding this admission, a Friday crucifixion does not fit scripturally at all (for me). May we grow in grace and knowledge to where we someday have this right. For it is the only sign given, and Christ said it (three days and three nights).

ERIC RASMUSSEN

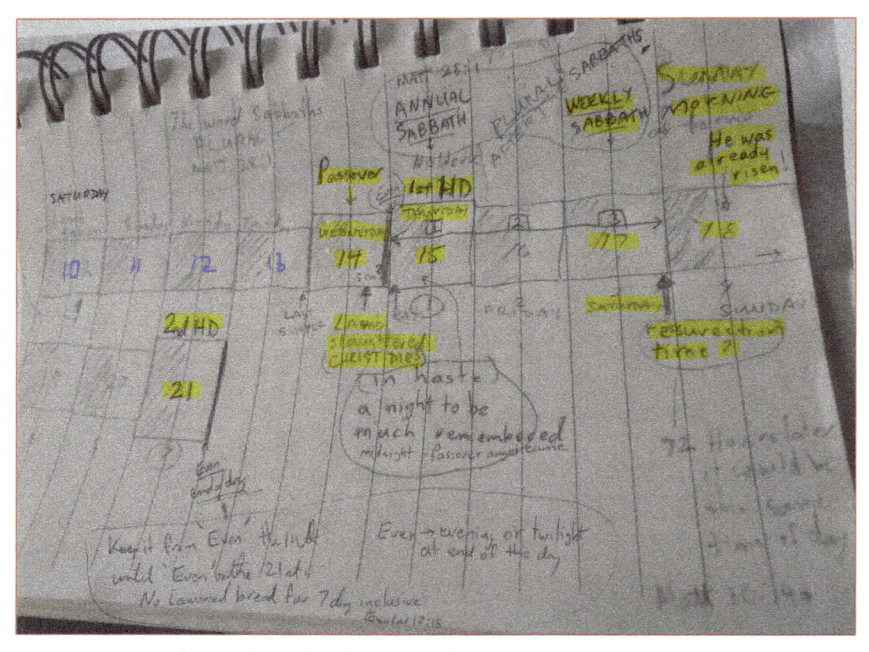

Commentary from F. LaGard Smith (daily Chronological Bible

Final Week—Wednesday

The preceding events of this final week appear to be accounted for by the Gospel writers within the clear context of either Sunday, Monday, or Tuesday, just as they have been presented. The exact timing of what happens after those events, however, appears less certain. John in particular touches only lightly upon the events between Jesus' triumphant entry and the so-called "last supper" which Jesus shares with his disciples. Referring to various public reactions during that time, John notes that, despite Jesus' teaching and miraculous works, there are still many people who either disbelieve or are afraid to acknowledge their belief. Then John records what is apparently Jesus' last public appeal before his subsequent arrest. As there is no evidence that these events occurred on any of the prior three days, they are set forth here as occurring on Wednesday, though that time frame is only speculative.

Of far greater significance at this point is the chronology related to the last supper, Jesus' crucifixion, and his subsequent resurrection. Traditionally the last supper is believed to have occurred on Thursday evening, followed by the crucifixion on Friday afternoon and the resurrection on Sunday morning. However, such reckoning raises at least two questions. First, in an action-packed final week, what reason is there to believe that there would be a whole day of either actual inactivity or activity which is left unrecorded? Second, and far more important—if Jesus is crucified on Friday afternoon and thereafter hurriedly put into the tomb, how can there be sufficient time to match Jesus' own prediction that he would remain in the tomb for three days and three nights before being resurrected? Even if one stretches imagination within the traditional time frame in order to find parts of three days, it is not possible to find three nights.

The resolution of both questions appears to be found in recognizing that the last supper took place on Wednesday evening, followed by the crucifixion and burial on Thursday. Acceptance of that assumption requires an understanding of the Passover, the Feast of Unleavened Bread, and the way in which the Jews reckon time. As for the reckoning of time, the Jewish day begins at sunset on the previous evening. This means, for example, that our Wednesday night is actually Thursday, and our Thursday night is actually Friday.

Passover is observed on the 14th day of the month of Nisan, corresponding to March-April. As noted earlier, Passover is observed in commemoration of the deliverance of the ancient Israelites from their Egyptian bondage. The name derives from the "passing over" of the Israelites when death came to the firstborn of each Egyptian family. As part of that same commemoration, Passover is followed by the seven-day Feast of Unleavened Bread, which reminds the Jews of their forefathers' flight from Egypt, during which time the Israelites ate unleavened bread only. (It is common among the Jews of Jesus' day to refer to both celebrations by only one name, either as "Passover" or as the "Feast of Unleavened Bread.") By God's direction (Leviticus 23), a lamb is to be slaughtered late on the 14th day (Passover) and the Passover meal eaten that evening, which would be the beginning of the 15th day, the first day of the Feast of Unleavened Bread. The entire 15th day is then to be observed as a special Sabbath, or high holy day, regardless of the day of the week on which it might fall in any given year. (If the 15th day is a Friday, then both that Friday and the next day, Saturday, are observed as Sabbaths.)

With that background the picture begins to come clear. Matthew, Mark, and Luke record the disciples' preparation for the Passover on the first day of the Feast of Unleavened Bread, on which the Passover lamb had to be sacrificed. That would place their preparations, then, at the beginning of the 14th day, which, of course, begins on the evening of the 13th day. (Among the preparations common on the evening of the 13th day is the removal of all leaven from the house.) Therefore it appears that the disciples assume they are preparing the upper room primarily for the special paschal meal which they expect to share with Jesus the following evening, and they apparently do not contemplate that the regular meal on the first night will in fact be their "last supper" with Jesus.

Although generally referring to the occasion as a part of the Passover celebration, Jesus seems to explain why it is important for him to eat with them on the night before the actual Passover meal. As will be seen, Jesus' words are: "I have eagerly desired to eat this Passover with you before I suffer. For I tell you, I will not eat it again until it finds fulfillment in the kingdom of God." In referring to his suffering, Jesus is obviously anticipating that his own sacrificial death will take place later that day, preventing him from participating in the actual Passover supper.

John's account eliminates any doubt that this supper occurred prior to the actual Passover meal. When Jesus tells Judas during the supper to do what he is about to do, some of the other disciples "thought Jesus was telling him to buy what was needed for the Feast." Furthermore, the Jews who have obtained Jesus' arrest will not enter Pilate's palace for fear that they will be ceremonially unclean, and therefore unable to eat the Passover. Most convincing is the fact that the day of Jesus' crucifixion is plainly stated to be "the day of Preparation of Passover Week"—the day on which the paschal

268

> NOVEMBER 13   1456
>
> lamb is slain for the Passover meal taken during the evening of that day.
>   The most meaningful result of moving away from the traditional time-frame is seeing how Jesus' crucifixion becomes the perfect "type" of the Passover Lamb. Under Hebrew law, the paschal lamb is chosen on the tenth day and then "kept up" until the 14th day, when it is sacrificed for the sins of the people. If Jesus' triumphant entry into Jerusalem is counted as the tenth day, Thursday would be the 14th day, and thus the day on which Jesus is crucified. Far more important than this possible parallel is the fact that Jesus, as the perfect Lamb of God, does not celebrate the Passover with some other ordinary sacrificial lamb, but rather becomes himself the Lamb who is slain—precisely at the appropriate hour!
>   There is therefore strong evidence that the last supper takes place on the evening prior to the Day of Preparation, which by modern reckoning would be Wednesday night. Proceeding upon that assumption, the events associated with this final Wednesday include not only Jesus' last public teaching, but also the account of Peter and John finding the upper room and making preparations for the Passover celebration.

15. Satan loosed

The God who would be man, Jesus, is the only man (both man and God) that could overcome Satan. We will know that <u>without a doubt</u>, and we will be full **of grace** and appreciation. *And nobody will ever say, given the chance, I could have done it.*

This time, where Satan is loosed after the millennium (perhaps one hundred years, I had heard in the past) proves no man can defeat him alone. We were all close to defeat until the Father said enough, and ==His will accomplished it.==

> *"In my vision at night I looked, and there before me was one like a **<u>son of man</u>, coming with the clouds of heaven**. He approached **the Ancient of Days** and was led into his presence. He was given authority, glory and sovereign power; **<u>all peoples, nations and men of every language</u> worshiped him**. His dominion is an everlasting dominion that will not pass away, and his kingdom is one that will never be destroyed.*
> *The Interpretation of the Dream*

"I, Daniel, was troubled in spirit, and the visions that passed through my mind disturbed me. I approached one of those standing there and asked him the true meaning of all this.

"So he told me and gave me the interpretation of these things: 'The four great beasts are four kingdoms that will rise from the earth. **But the saints of the Most High will receive the kingdom and will possess it forever**—*yes, for ever and ever.'*

whatever was left. I also wanted to know about the ten horns on its head and about the other horn that came up, before which three of them fell—the horn that looked more imposing than the others and that had eyes and a mouth that spoke boastfully. **As I watched, this horn was waging war against the saints and defeating them, until the Ancient of Days came and pronounced judgment in favor of the saints of the Most High**, *and the time came when they possessed the kingdom. (Daniel 7:13–22, NIV)*

I found the article below on the Internet. I wanted to share the importance of what it says about man earning salvation.

The events prove it. We are saved by grace alone. I think it extends and applies for all mankind of any generation. I say that because I was once pretty sure that if I just kept God's law and did my part of the bargain, I would make it into the kingdom. I now realize I was conceived in sin and must have God change me from that by the indwelling presence of Christ's and the Father's Spirits coming into me.

Against you, you only, have I sinned
and done what is evil in your sight,
so that you are proved right when you speak
and justified when you judge.
Surely I was sinful at birth,
sinful from the time my mother conceived me.

SPACE THE TRUE FRONTIER!

Surely you desire truth in the inner parts;
you teach me wisdom in the inmost place.
Cleanse me with hyssop, and I will be clean;
wash me, and I will be whiter than snow.
Let me hear joy and gladness;
let the bones you have crushed rejoice.
(Psalms 51:4–8, NIV)

Hide your face from my sins
and blot out all my iniquity.
Create in me a pure heart, O God,
and renew a steadfast spirit within me.
Do not cast me from your presence
or take your Holy Spirit from me.
Restore to me the joy of your salvation
and grant me a willing spirit, to sustain me.
(Psalms 51:9–12, NIV)

We do all have a choice, and for some, the choice happens unexpectedly after the millennium.

ERIC RASMUSSEN

Satan is Let Loose One More Time

After the 1000 years have ended in the Millennium Kingdom, two additional things must happen before we get our new heaven and new earth. These two events will seal the fate of Satan and all of the unsaved people who have ever lived once and for all.

This first event is almost unbelievable. When Jesus comes back for the second time to set up His Millennium Kingdom for a thousand years, the Bible tells us that Satan is chained up in the bottomless pit during those thousand years. He will no longer be free to tempt or torment humans during the Millennium Kingdom.

However, after the thousand years have ended in the Millennium Kingdom, Satan is let loose for one last time to come upon our earth to try and cause one more round of trouble.

The original King James Version tells us that Satan will be let loose for a **"little season."** The New King James Version says he will be let loose for a **"little while."** The Bible does not tell us how many years Satan will be allowed to run free again. All that we can surmise from the above words is that he will only be free to roam for a very short period of time.

As soon as he is released from the bottomless pit, Satan will waste no time in doing what he does best. The Bible says that he will then go out to deceive the nations once again. He will be allowed to literally go to the 4 corners of the globe to get as many people as he possibly can to do his evil bidding once again.

The Bible says that he will be targeting all of the people who are wanting to rebel against God from the Millennium Kingdom. He will get them to surround the camps of the saints and the beloved city, which is probably Jerusalem, and then get them to try and attack God's people and holy city.

However, the Bible tells us they will not even be able to get to the point of actually launching any type of attack against Jerusalem or any of God's people.

They get to the point where they are surrounding Jerusalem, but before they can even attack, God Himself will literally devour them by sending down fire from heaven. God will literally take them out right there on the spot! There will be no physical battles like there will be in the battle of Armageddon. No blood will be spilled with the passing of this event.

And then the Bible tells us that Satan will then be thrown into the Lake of Fire and Brimstone where the Antichrist and False Prophet had been thrown in earlier at the beginning of the Millennium Kingdom.

SPACE THE TRUE FRONTIER!

> Once this is done, the Bible says they will remain in this place forever and ever! We will thus never hear from the Devil again. God will have finally taken out this blasphemous creature from His sight and our sight for all of eternity!

If we have just had perfect peace for a thousand years in the Millennium Kingdom, why would God allow the devil to be set free to cause one more round of trouble? Here is my opinion as to why.

In the Millennium Kingdom, the humans who survive the events of the Tribulation will still be procreating and bringing more children into our world.

These children will then be raised up in the Lord and all of His ways. They will have never known demonic temptation since Satan and all of his demons will have been chained up and out of their sight during the entire length of the Millennium Kingdom.

If Satan is let loose and manages to gather a certain amount of people to try and attack Jerusalem, then this means that some of these grown up children are going to rebel and go against God Almighty Himself in this last event.

The fact that Satan will be able to easily dupe some of them into actually doing this tells me that some of these people had to have been restless and not totally committed to serving the Lord in the Millennium Kingdom. Otherwise, they would have never been so easily tempted by Satan to commit this one, last, final act of rebellion against God's people and His holy city of Jerusalem.

I believe the main reason God will let Satan loose for one more time is to prune out some of these bad people who won't stay loyal and committed to serving Him.

These people will not be able to start any kind of trouble against God in the Millennium Kingdom because Satan is not around yet, people no longer war with one another, and we will have perfect law and order in this new environment since Jesus will now be in full control of everything.

I believe God will set these people up. He will allow Satan to roam free one more time so he can tempt these people into going against Him and His people. And right at the point where these rebellious ones are getting ready to encircle Jerusalem to attack it, God will literally destroy them right there on the spot.

What this tells us is that there will still be some people who will rebel and go against God – even in the most perfect God-environment imaginable!

These people will have all been born into and raised up in this Millennium Kingdom without any type of bad influence from Satan, his demons, and any other bad and unsaved people, since God had already thrown all of them into Hades for the entire length of the Millennium Kingdom.

These people will have also been raised up being taught who the real God is, along with being taught all of His awesome ways. There will be no other gods or false religions to confuse or seduce them.

And even in this most perfect scenario imaginable, there is obviously going to be a group of these people who will still choose to rebel against God and try to overthrow Him by attacking His holy city of Jerusalem once Satan is let loose to try and set all of this up.

When you first examine all of the above, you have to ask yourself, how could a group of humans be so stupid to think they can overthrow or go against God Almighty Himself, especially since they have probably been taught the entire history of our world during the Millennium Kingdom?

How could these people go against God when they have been raised up in the most perfect God-environment imaginable?

To answer this, all you have to do is look at what happened to Satan and all of the fallen angels. They were all initially living up in heaven with the Lord with no negative influences until Satan started to malfunction through pride and conceit.

They too were living in the most perfect God-environment imaginable, and possibly up to one-third of them decided with their own free wills to rebel against God Himself in this perfect heavenly environment.

So just as a certain of amount of the angels decided to rebel and go against God Himself in a perfect heavenly environment, so too are a certain number of human beings going to do the exact same thing at the end of the Millennium Kingdom. And once again, God will allow Satan to be the one to tempt and cause these rebellious ones to fall.

Bottom line – God knows there are going to be a certain number of people who will still try and come against Him, even in this perfect Millennium Kingdom. He is simply going to draw them out, expose them, and then prune them out by allowing Satan to deceive them like he did with Adam and Eve in the Garden of Eden.

Then once all of this has been accomplished, you will have nothing but the good apples left that will finally inherit God's best for all of humanity – The New Heaven and the New Earth.

Now here are the two main verses from our Bible that will give us all of the above information.

The Scripture Verses

1. This first verse will tell us that Satan will be let loose for a little while after he is released from the bottomless pit.

"Then I saw an angel coming down from heaven, having the key to the bottomless pit and a great chain in his hand. He laid hold of the dragon, that serpent of old, who is the Devil and Satan, and bound him for a thousand years, and he cast him into the bottomless pit, and shut him up, and set a seal on him, so that he should deceive the nations no more till the thousand years were finished. But after these things he must be released for a little while." (Revelation 20:1-3)

SPACE THE TRUE FRONTIER!

2. This next verse will tell us what happens after he is released from the bottomless pit.

"Now when the thousand years have expired, Satan will be released from his prison and will go out to deceive the nations which are in the four corners of the earth, Gog and Magog, to gather them together to battle, whose number is as the sand of the sea. They went up on the breadth of the earth and surrounded the camp of the saints and the beloved city. And fire came down from God out of heaven and devoured them. And the devil, who deceived them, was cast into the lake of fire and brimstone where the beast and the false prophet are. And they will be tormented day and night forever and ever." (Revelation 20:7-10)

Now before God gives us our final reward with the New Heaven and New Earth, there is one more really bad event that must take place. This event is called the Great White Throne Judgment. This event will determine the eternal fate of all of the unsaved people who have ever lived. This event will discussed in the next article.

Source: http://www.bible-knowledge.com/satan-let-loose/

16. The name Lucifer should not be in the Bible.

"Lucifer Means Lightbringer," Lucifermeanslightbringer, last modified July 3, 2016, https://lucifermeanslightbringer.com/2016/07/03/lucifer-means-lightbringer-2/.

> Most people are familiar with the concept of Lucifer as being another name for Satan, the devil of the Christian religion. However, this is a recent association and the concept of Lucifer goes much further back into history. The Hebrew word translated as Lucifer, הֵילֵל בֶּן־שָׁחַר (Helel ben Shaḥar), is simply the Hebrew word for Venus as the Morningstar - just like the Latin word 'lucifer,' it also translates to "light-bringer," "shining one," "son of the morning," etc. Helel ben Shaḥar actually only appears once in the Old Testament of Bible (the New Testament, including Revelation, was written in Greek), and most scholars agree that it was being used metaphorically to refer to the Babylonian king Nebuchadnezzar II, who conquered Jerusalem, but then suffered great setbacks - he rose and fell like Venus, in other words. The word 'lucifer' was only later associated with Satan as a fallen angel by Pope Gregory the Great (540-604 AD), an idea which was then popularized by John Milton in 'Paradise Lost.' Here's the verse:
>
> Isaiah 14:12: "How art thou fallen from heaven, O Lucifer, son of the morning! How art thou cut down to the ground, thou which didst weaken the nations!"
>
> Saying that Satan's name is "Lucifer" is no different than saying his name is "Phosphorus," except that 'Lucifer' is much catchier. As I mentioned before, Lucifer is a Latin word much older than Pope Gregory or John Milton, and it simply refers to the planet Venus when it appears as the Morningstar. Pope Gregory wasn't mad, however - in fact, he had very good reason to make the association between Satan and Venus. The fable of a high angel who challenges god and is thrown out of heaven to become the king of hell is one interpretation of the universal mythological archetype known as the Morningstar deity.
>
> But then again, so is the story of Jesus Christ, who is named the Morningstar in the New Testament book of Revelation, chapter 22, verse 16:
>
> "I, Jesus, have sent my angel to give you this testimony for the churches. I am the Root and the Offspring of David, and the bright Morning Star."

17. Understanding Parsing, in the ISA program.

The next three pictures show how the parsing is done in the Interlinear Scripture Analyzer software. Note especially where gender is displayed.

ΠΑΡΑΚΛΗΤΟΣ	ΤΟ	ΠΝΕΥΜΑ
paraklEtos	to	pneuma
BESIDE-CALLer	THE	spirit
consoler		
n_ Nom Sg m	t_ Nom Sg n	n_ Nom Sg n

SPACE THE TRUE FRONTIER!

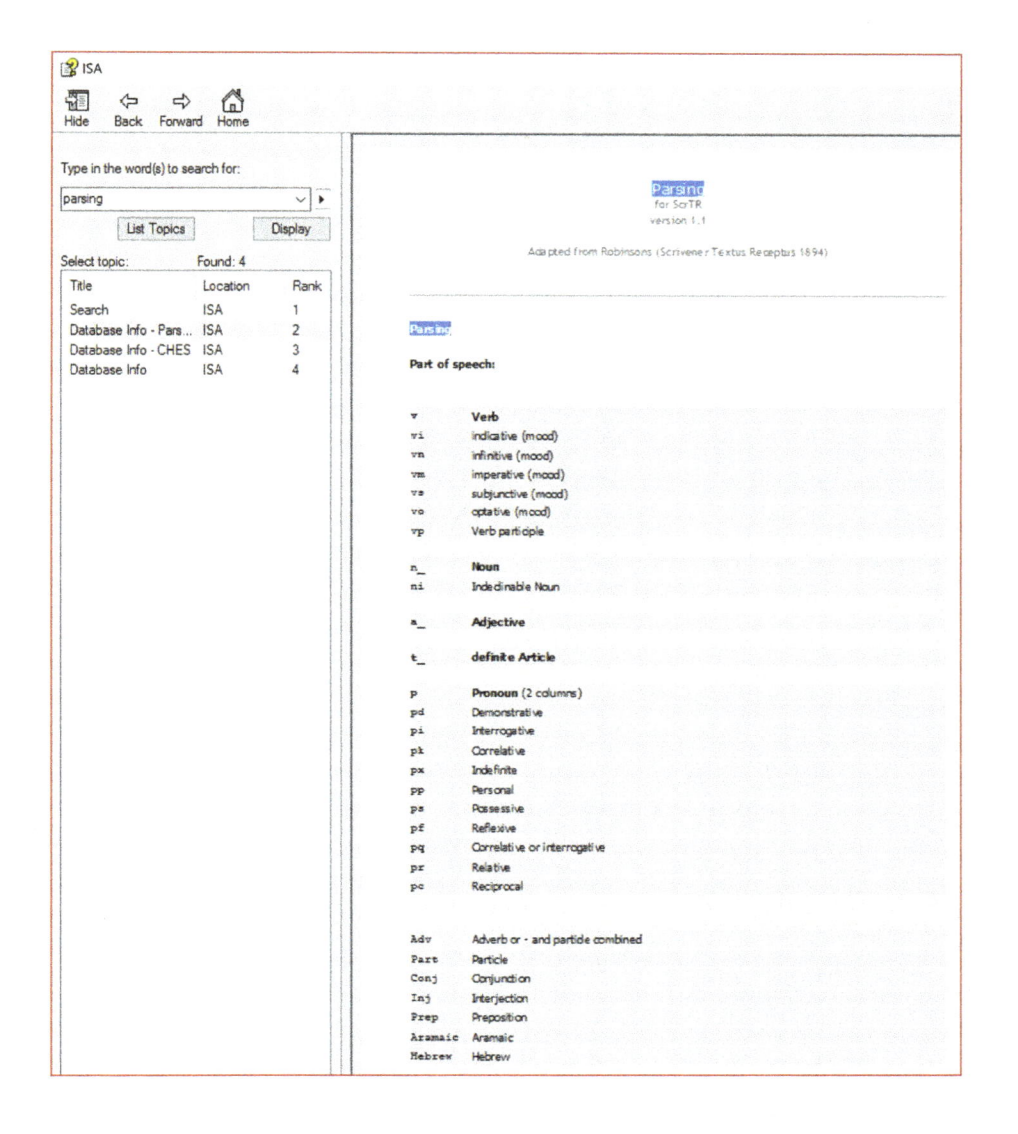

Parsing
for ScrTR
version 1.1

Adapted from Robinsons (Scrivener Textus Receptus 1894)

Parsing

Part of speech:

v	**Verb**
vi	indicative (mood)
vn	infinitive (mood)
vm	imperative (mood)
vs	subjunctive (mood)
vo	optative (mood)
vp	Verb participle

n_	**Noun**
ni	Indeclinable Noun

a_	**Adjective**

t_	**definite Article**

p	**Pronoun** (2 columns)
pd	Demonstrative
pi	Interrogative
pk	Correlative
px	Indefinite
pp	Personal
ps	Possessive
pf	Reflexive
pq	Correlative or interrogative
pr	Relative
pc	Reciprocal

Adv	Adverb or - and particle combined
Part	Particle
Conj	Conjunction
Inj	Interjection
Prep	Preposition
Aramaic	Aramaic
Hebrew	Hebrew

277

ISA

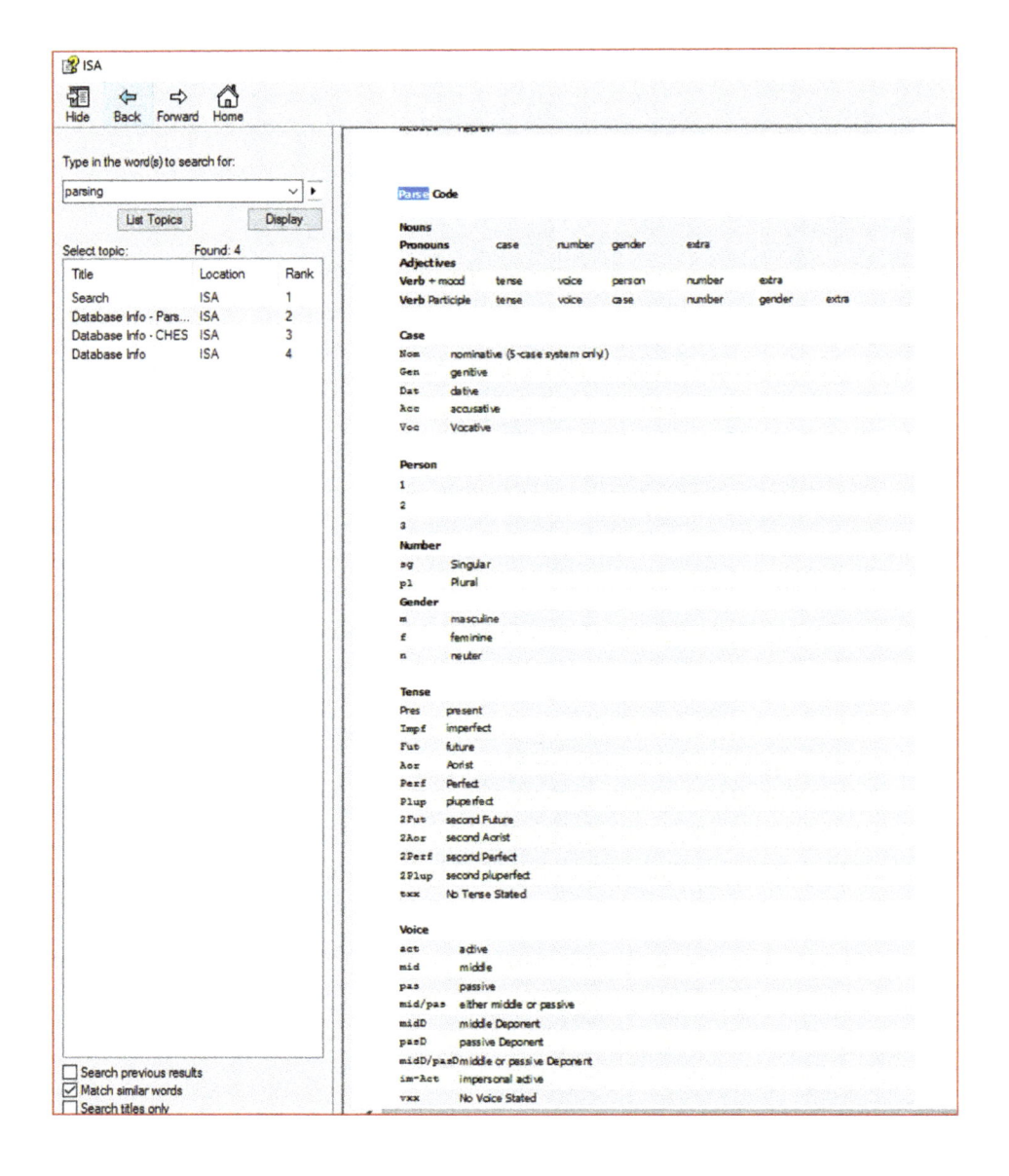

Type in the word(s) to search for:

parsing

[List Topics]   [Display]

Select topic: Found: 4

Title	Location	Rank
Search	ISA	1
Database Info - Pars...	ISA	2
Database Info - CHES	ISA	3
Database Info	ISA	4

☐ Search previous results
☑ Match similar words
☐ Search titles only

Parse Code

Nouns
Pronouns   case   number   gender   extra
Adjectives
Verb + mood   tense   voice   person   number   extra
Verb Participle tense   voice   case   number   gender   extra

Case
Nom   nominative (5-case system only)
Gen   genitive
Dat   dative
Acc   accusative
Voc   Vocative

Person
1
2
3

Number
sg   Singular
pl   Plural

Gender
m   masculine
f   feminine
n   neuter

Tense
Pres   present
Impf   imperfect
Fut   future
Aor   Aorist
Perf   Perfect
Plup   pluperfect
2Fut   second Future
2Aor   second Aorist
2Perf   second Perfect
2Plup   second pluperfect
txx   No Tense Stated

Voice
act   active
mid   middle
pas   passive
mid/pas   either middle or passive
midD   middle Deponent
pasD   passive Deponent
midD/pasD middle or passive Deponent
im-Act   impersonal active
vxx   No Voice Stated

278

SPACE THE TRUE FRONTIER!

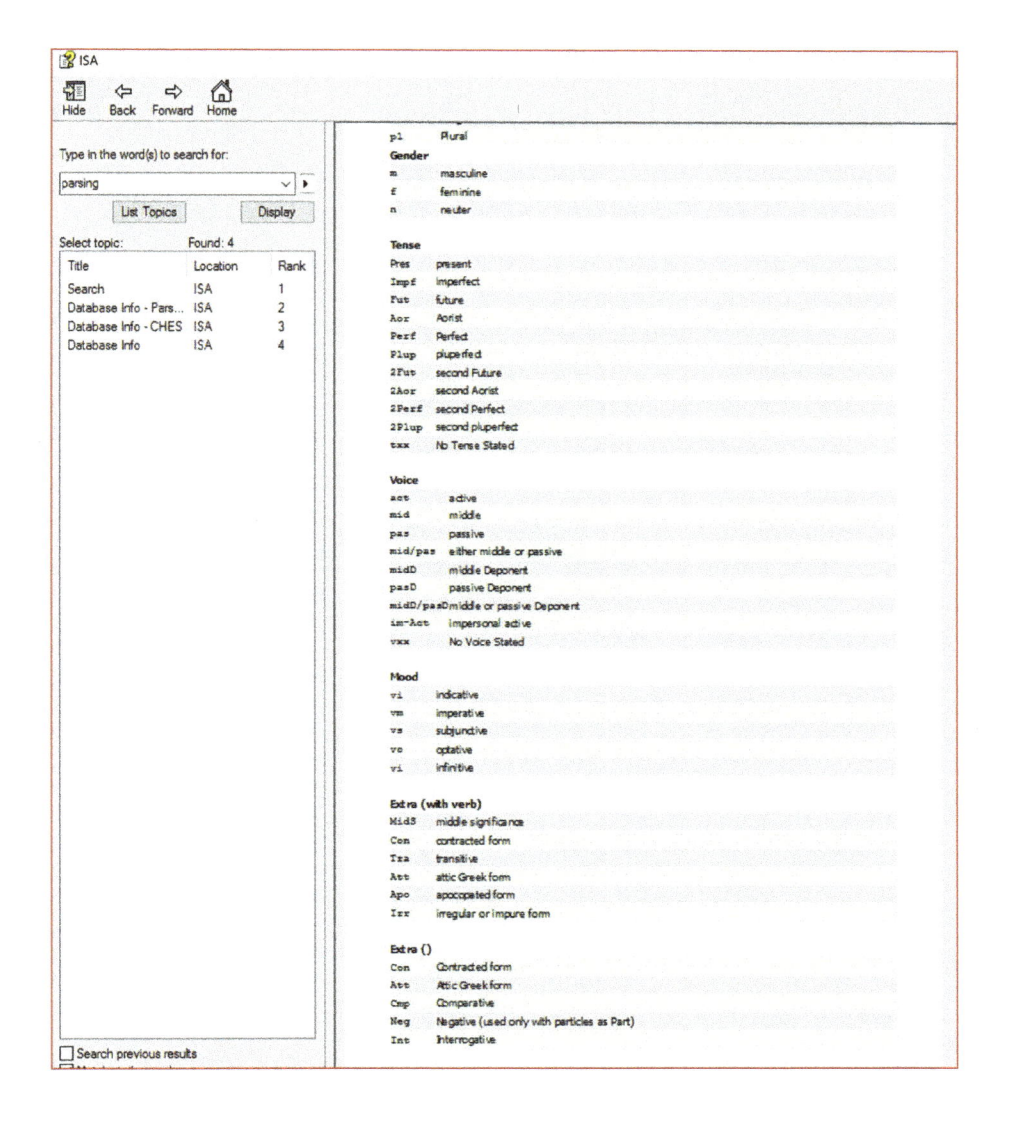

ISA

Type in the word(s) to search for:
parsing

[List Topics] [Display]

Select topic: Found: 4

Title	Location	Rank
Search	ISA	1
Database Info - Pars...	ISA	2
Database Info - CHES	ISA	3
Database Info	ISA	4

☐ Search previous results

```
pl         Plural
Gender
m          masculine
f          feminine
n          neuter

Tense
Pres       present
Impf       imperfect
Fut        future
Aor        Aorist
Perf       Perfect
Plup       pluperfect
2Fut       second Future
2Aor       second Aorist
2Perf      second Perfect
2Plup      second pluperfect
txx        No Tense Stated

Voice
act        active
mid        middle
pas        passive
mid/pas    either middle or passive
midD       middle Deponent
pasD       passive Deponent
midD/pasD  middle or passive Deponent
im-Act     impersonal active
vxx        No Voice Stated

Mood
vi         indicative
vm         imperative
vs         subjunctive
vo         optative
vi         infinitive

Extra (with verb)
MidS       middle significance
Con        contracted form
Tra        transitive
Att        attic Greek form
Apo        apocopated form
Irr        irregular or impure form

Extra ()
Con        Contracted form
Att        Attic Greek form
Cmp        Comparative
Neg        Negative (used only with particles as Part)
Int        Interrogative
```

279

18. Was Genesis a re-creation?

"Tohu Wabohu: The Opposite of God's Creational Intent," Academia.edu - Share Research, accessed November 30, 2017, http://www.academia.edu/4255454/Tohu_Wabohu_The_Opposite_of_Gods_Creational_Intent.

Tohu Wabohu and the Negative Condition of the Land

In the discussion of the term tohu wabohu it will be seen that in every other occurrence of that term in Scripture (including the word tohu when used alone), the context shows that judgment has been involved in some way. It would therefore be reasonable to attribute a context of judgment to Genesis 1:2 as well. If that is admitted as a valid hypothesis, then saying that "the land was" tohu wabohu (as a result of judgment) means the same as "the land had become" tohu wabohu. In either case, it had experienced a change from conditions hospitable to life to uninhabitable conditions. The author of Genesis shows in the first chapter how God goes about correcting these conditions that are contrary to his will. He does this by emphasizing a definite pattern in the creationstory that is given to enable God's people to imitate him in overcoming evil with good. The negative condition of the land preceding God's creative, corrective work is evident in Bruce Waltke's detailed comparison of Genesis 1:1, 2 to Genesis 2:4-7. (1975:132:225 ff.) (John Sailhamer considers that "the Genesis 1 and 2 narratives are about the same events and have the same setting. What we see God doing in Genesis 2 is merely another perspective on what He does in Genesis 1." [1996: 51]) Each of these passages has an introductory statement that summarizes the rest of the chapter, followed by a circumstantial clause that modifies the upcoming verse. This second element in each case follows the non-usual, emphatic pattern "waw + noun + verb (hayah)" describing a negative state before creation. Finally the introduction to each creation account brings in the main clause that uses the normal Hebrew verb/subject pattern "waw consecutive +prefixed conjugation form describing the creation." (Waltke 1975: 132:225) The editors of the Net Bible give a somewhat simpler summary: "This literary structure [of Genesis 1:1, 2] is paralleled in the second portion of the book: Gen 2:4 provides the title or summary of what follows, 2: 5-6 use disjunctive clause structures to give background information for the following narrative, and 2:7 begins the narrative with the vav consecutive attached to a prefixed verbal form."

Additional note: according to F. F. Bruce, the first three verses of Genesis run as follows in Hebrew:

(1) Be-reshith bara Elohim eth ha-shamayim we-eth ha-arets: (2) we-ha-arets hayethah ==tohu wa-bohu== we-choshekh al-pnê tehôm we-rûach Elohim merachepheth al-pnê ha-mayim: (3) wayyômer Elohim "yehi ôr" wa-yehi ôr.

The question before us is whether (a) "ver. 2 implies the occurrence of some change of catastrophic order subsequent to creation, and that the earth had become 'without form and void,'" or (b) "ver. 2 merely defines the condition of the earth at its creation." The terms of reference prescribe a strictly linguistic discussion, excluding all considerations of the relation between these verses and theological or natural science.

Note: that FF Bruce does not agree that there was a recreation—see his entire article (F. F. Bruce, *And the Earth was Without Form and Void: An Enquiry into the Exact Meaning of Genesis 1, 2*, Journal of the Transactions of the Victoria Institute 78 (1946): 21–37, found in "Making Biblical Scholarship Accessible Since 2001, https://biblicalstudies.org.uk/pdf/jtvi/without-form_bruce.pdf.)

19. Cosmos

==It is interesting to consider the "order" of the universe is the source of the meaning of its very name, *cosmos*.== It must have taken a lot of energy to put order into all that, and for all of that to be wound up and exist, especially because we know it is now unwinding as entropy increases.

I think common sense should lead us to realize that the source of such order couldn't be a Higgs boson field (which is sciences source for all matter, including the so-called God particle, Higgs boson particle discovered in 2012) but would obviously need to be something of unimaginable power and intelligence (God).

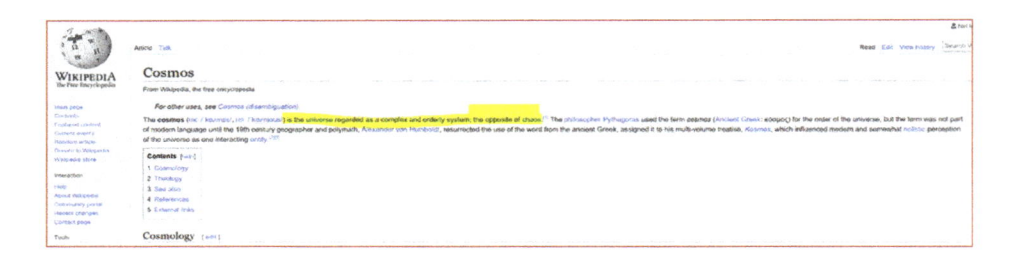

20. "The Amazing Name Israel: Meaning and Etymology," Abarim Publications, accessed November 30, 2017, http://www.abarim-publications.com/Meaning/Israel.html#.WW6APGwUhoc.

SPACE THE TRUE FRONTIER!

Etymology of the name Israel

The meaning of the name Israel is not clear, but yet it's huge. The meaning of Israel is not singular and distinct, but consists of many nuances and facets and bulges with theological significance.

Judging from Genesis 32:28, the form ישראל (Israel) appears to be a compilation of two elements. ==The first one is the noun אל (El), the common abbreviation of Elohim, meaning God:==

Abarim Publications' online Biblical Hebrew Dictionary

אל אלה

In names, the segment אל ('*el*), usually refers to אלהים ('*elohim*), that is Elohim, or God, also known as אלה ('*eloah*). In English, the words 'God' and 'god' are strictly reserved to refer to deity but in Hebrew the words אל ('*l*) and אלה ('*lh*) are far more common. Consider the following:

אל

- אל ('*al*), which is the Hebrew transliteration of the Arabic article that survives in English in words like alcohol and algebra. There are some words in the Hebrew Bible that are transliterations of Arabic words, which contain this article.
- אל ('*al*), particle of negation; not, no, neither.
- אל ('*el*) preposition that expresses motion towards someone or something; unto, into, besides, in reference to.
- אל ('*el*), which is a truncated form of אלה ('*eleh*), meaning these (see below).

אלה

- אלה ('*eleh*), meaning these. Follow the link to read our article on this and the next three words
- אלה ('*ala*), to swear; derivative אלה ('*ala*) means oath.
- אלה ('*ala*), to wail.
- אלה ('*alla*), oak, from the assumed and unused root אלל ('*ll*). Follow the link to read more on these and the next words
- אלה ('*ela*), terebinth, from the root אול ('*wl*).

Associated Biblical names

♂	Abdeel	עבדאל	♂⚪	Ariel	אריאל	⚪	Beth-arbel	בית ארבאל
♂	Abdiel	עבדיאל	♂	Asahel	עשהאל	⚪	Bethel	בית־אל
♂	Abiel	אביאל			עשה־אל	♂⚪	Bethuel	בתואל
♂	Abimael	אבימאל	♂	Asarel	אשראל	♂●	Bezalel	בצלאל
♂	Adbeel	אדבאל	♂	Asiel	עשיאל	♂	Daniel	דניאל
♂	Adiel	עדיאל	♂	Asriel	אשריאל			דנאל
♂	Adriel	עדריאל	♂	Azarel	עזראל	♂	Deuel	דעואל
♂	Almodad	אלמודד	♂♠⊘	Azazel	עזאזל	♂☼✱	El	אל
♂	Ammiel	עמיאל	♂	Azriel	עזריאל	♂✱	El-berith	אל ברית
♂	Areli	אראלי	♂	Barachel	ברכאל	⚪☼	El-bethel	אל בית־אל

♂	Eldaah	אלדעה	♂②	Eli	עלי	♂	Eliathah	אליאתה
♂	Eldad	אלדד			אלי			אליותה
♂	Elead	אלעד	♂	Eliab	אליאב	♂	Elidad	אלידד
♂	Eleadah	אלעדה	♂	Eliada	אלידע	♂	Eliehoenai	אליהועיני
⚪	Elealeh	אלעלה	♂	Eliahba	אליחבא	♂	Eliel	אליאל
		אלעלא	♂☗	Eliakim	אליקים	♂	Elienai	אליעני
	Eleasah	אלעשה	♂	Eliam	אליעם	♂	Eliezer	אליעזר
♂	Eleazar	אלעזר	♂	Eliasaph	אליסף	♂	Elihoreph	אליחרף
⚪	El-Elohe-Israel	אל אלהי ישראל	♂	Eliashib	אלישיב	♂	Elihu	אליהו
♂	Elhanan	אלחנן						אליהוא

♂	Elijah	אליה	♂⚪	Elishah	אלישה	♂☼	El-kana	אל קנא
♂	Elika	אליקא	♂	Elishama	אלישמע			אל קנוא
♂	Elimelech	אלימלך	♂	Elishaphat	אלישפט	♂	Elkanah	אלקנה
♂	Elioenai	אליועיני	♀	Elisheba	אלישבע	♂	Elkoshite	אלקשי
		אליועני		Elishua	אלישוע	⚪	Ellasar	אלסר
♂	Eliphal	אליפל	♂	Eliud	Ελιουδ	♂	Elmadam	αλμωδαμ
♂	Eliphaz	אליפז	♀	Elizabeth	Ελισαβετ	♂	Elnaam	אלנעם
♂	Eliphelehu	אליפלהו	♂	Elizaphan	אליצפן	♂	Elnathan	אלנתן
♂	Eliphelet	אליפלט	♂	Elizur	אליצור	♂∩☼	Elohim	אלהים
♂	Elisha	אלישע				♂	Elpaal	אלפעל

⚪	Elteke	אלתקה	♂	Gaddiel	גדיאל	⚪	Iphtahel	יפתח־אל
		אלתקא	♂♠	Gamaliel	גמליאל	♂	Ishmael	ישמעאל
⚪	Eltekon	אלתקן	♂	Hammuel	חמואל	♂⚪	Israel	ישראל
♂	Eltolad	אלתולד	♂⚪	Hananel	חננאל	♂	Ithiel	איתיאל
♂	Eluzai	אלעוזי	♂	Hanniel	חניאל	♂	Jaasiel	יעשיאל
♂	Elymas	Ελυμας	♂☗	Hazael	חזאל	♂	Jaaziel	יעזיאל
♂	Elzabad	אלזבד			חזהאל	⚪	Jabneel	יבנאל
♂	Ezekiel	יחזקאל	♂	Haziel	חזיאל	♂	Jahaziel	יחזיאל
♂	Gabriel	גבריאל	♂	Hiel	חיאל	♂	Jahdiel	יחדיאל
♀	Gabriela	גבריאלה	♂☼	Immanuel	עמנו אל	♂	Jahleel	יחלאל

♂	Jahzeel	יחצאל	♂	Asharelah-Jesharelah	אשראלה	♂♀	Mehetabel	מהוטבאל
♂	Jathniel	יתניאל			ישראלה	♂	Mehujael	מחויאל
♂	Jediael	ידיעאל	♂⚪	Jezreel	יזרעאל			מחייאל
♂	Jehallelel	יהללאל		Joel	יואל	♂	Methushael	מתושאל
♂	Jehiel	יחיאל	♂	Kabzeel	קבצאל	♂	Michael	מיכאל
		יחואל	♂	Kemuel	קמואל	♀	Michal	מיכל
♂	Jekuthiel	יקותיאל		Lael	לאל	♂	Migdal-el	מגדל־אל
♂	Jerahmeel	ירחמאל	♂	Lazarus	Λαζαρος	♂	Nathanael	Ναθαναηλ
♂	Jeriel	יריאל	♂	Mahalalel	מהללאל	♂	Nethanel	נתנאל
⚪	Jeruel	ירואל	♂	Malchiel	מלכיאל	♂	Paltiel	פלטיאל

♂⚪	Peniel	פניאל	♂	Shebuel	שבואל			
♂⚪	Penuel	פנואל			שובאל			
♂	Pethuel	פתואל			שבאל			
♂	Phanuel	Φανουηλ	♂	Shelumiel	שלמיאל			
♂	Raphael	רפאל	♂	Tabeel	טבאל			
♂	Reuel	רעואל	⚪	Taralah	תראלה			
♂	Samuel	שמואל	♂	Uriel	אוריאל			
♂	Shealtiel	שאלתיאל	♂	Uzziel	עזיאל			
		שלתיאל	♂	Zabdiel	זבדיאל			
			♂	Zuriel	צוריאל			

SPACE THE TRUE FRONTIER!

The second part of our name appears to be related to the verb שרה I (sara I):

Abarim Publications' online Biblical Hebrew Dictionary

שרה שרר

The forms שרה (srh) and שרר (srr) are part of an enormous cluster of words, some of which are obviously related. Note that the difference between שׂ (sin; dot to the left, probably pronounced similar to our letter s) and שׁ (shin; dot to the right, probably pronounced as sh) is an interpretation made by the Masoretes more than a thousand years after the text of the Bible was written. The Biblical authors used only the letter ש (s; no dot; pronunciation probably somewhere in between s and sh):

שׂרר

The basic meaning of the root שרר (srr) is unclear but a similar root-verb in Assyrian, sararu means to rise in splendor (of the sun, for instance). BDB Theological Dictionary, however, deems to connection dubious. The Bible reflects this root in two closely related nouns and a denominative verb:

- The masculine noun שר (sar), meaning chief or ruler. This common noun mostly denotes a social structure's sub-chief, like a clan head (Numbers 21:18) or regional ruler (Judges 9:30). In a few occasions the שר (sar) is an angelic captain (Joshua 5:14, Daniel 10:13).
- The feminine equivalent שרה (sara), denoting a princess or noble lady (Judges 5:29, Isaiah 49:23).
- The denominative verb שרר (sarar), meaning to be or act as a שר (sar), or in short: to rule or exercise dominion (Isaiah 32:1, Esther 1:22).

שׁרר

The root שרר (srr) appears to be related to words in cognate languages that have to do with firmness and hardness and even to be substantial and truthful. Perhaps it's a coincidence but these qualities are obviously those of a righteous ruler. The usages of this root in the Bible reveal this root's secondary charge of centrality, also a characteristic of a king or ruler:

- The masculine noun שר (shor) meaning umbilical cord (Proverbs 3:8, Ezekiel 16:4).
- The feminine noun שרה (shera), meaning bracelet (Genesis 24:22, Isaiah 3:19).
- The masculine noun שריר (sharir), apparently denoting a sinew or muscle (Job 40:16 only).
- The feminine noun שרירות (sherirut) or שרירת (sherirut), meaning firmness in a negative sense: stubbornness. This noun is used always in a context with the noun לב (leb), meaning heart, the central-most organ and the Biblical seat of the mind.

שׂרה I

The meaning of the verb שרה (sara I) is uncertain and explained in many ways, chiefly because it is limited to contexts which discuss the struggle of Jacob with the Angel of YHWH (Genesis 32:29 and Hosea 12:4 only), insinuating that where our language uses the common verb 'struggle,' the Hebrew uses a word that is specifically reserved for a certain action: the action of struggling with God.

BDB Theological Dictionary reports for שרה (sara) the Arabic cognate of to persist, persevere and interprets our verb as such. HAW Theological Wordbook of the Old Testament believes our verb to mean to contend or have power.

Perhaps a Hebrew audience would have viewed this enigmatic verb as having to do with the previous roots (containing words that have to do with royalty), possibly concluding that Jacob didn't simply stand up to a celestial bully, but rather that the angel saw in Jacob a worthy national ruler. The struggle of Jacob with the angel was not so much a bout between two hulks, but rather an international power struggle that resulted in an earth-heaven federation.

שׂרה II

Linguists insist that the form שרה (srh) must be split into two separate roots, but why is not very clear. In the Bible the assumed root שרה (srh) is only reflected in the masculine noun משרה (misra), which only occurs in the famous Messianic passage of Isaiah 9:6: "... and the *government* will be upon His shoulders". This is obviously not very far removed from the roots שרר (srr).

285

שׂרה

The following cluster of roots that are all spelled שׂרה (*shrh*) appear to reflect attributes of the royal office:

שׂרה I

The verb שׂרה (*shara* I) means to release or let loose. It's used two times in the Bible. In Job 37:3, YHWH releases thunder and lightening from the heavens, in a passionate report that celebrates the Lord as the ruler of the earth. In Jeremiah 15:11, the Lord is portrayed as a military leader who promises to release Jeremiah from the enemy.

שׂרה II

The root-verb שׂרה (*shrh* II) doesn't occur in the Bible but in cognate languages it exists with the meaning of to be moist. In the Bible only one derivative exists, namely the feminine noun משׁרה (*mishra*), denoting the juice of grapes. This noun occurs only once, in Numbers 6:3, where the juice of grapes is distinguished from fresh grapes or dried grapes.

שׂרה III

The root שׂרה (*shrh* III) also doesn't occur in the Bible. Its sole derivative is the feminine noun שׂריה (*shirya*), which denotes some kind of weapon, most likely a ballistic one; perhaps a lance or javelin. It occurs only once, in Job 41:26.

שׂרה IV

Root שׂרה (*shrh* IV) is also not used, and only one derivative remains: the masculine noun שׁריון (*shiryon*) or שׁרין (*shiryan*), meaning body armor (1 Samuel 17:5, 1 Kings 22:34).

Associated Biblical names

♂♫	Ahishar	אחישׁר
○	El-Elohe-Israel	אל אלהי ישׂראל
♂○	Israel	ישׂראל
♀	Sarah	שׂרה
♀	Sarai	שׂרי
♀	Serah	שׂרח
♂	Seraiah	שׂריה
♂☾	Shamsherai	שׁמשׁרי
♂	Sharar	שׁרר
♀○●	Sharon	שׁרון

However, even though Genesis 32:28 uses the enigmatic verb שׂרה — which is assumed to mean to struggle but which might something else entirely — it's by no means certain that this verb is etymologically linked to our name Israel. When we say, "we named him Bob because that seemed like a good idea," we certainly don't mean to say that the name Bob means "good idea".

The first part of the name Israel looks a lot like the verb שׂרה that explains this name, but this apparent link is possibly a mere case of word-play. In fact, the name Israel may have more to do with the verb ישׁר (*yashar*), meaning to be upright. Note that the difference between the letter שׂ (*sin*) as found in the name ישׂראל (Israel) and the letter שׁ (*shin*) as found in the verb ישׁר (*yashar*) didn't exist in Biblical times and as it was invented more than a thousand years after the Bible was written;

SPACE THE TRUE FRONTIER!

Abarim Publications' online Biblical Hebrew Dictionary

אשר ישר

The form אשר ('sr) occurs in two different ways: There's the verbal root אשר ('ashar), which indicates progression, and there's the particle אשר ('asher) that indicates relation. Whether the two are etymologically related isn't clear, although there seems to be an obvious intuitive connection. And then there is the verb ישר (yashar), which appears to be etymologically related and certainly is so in meaning:

אשר

The root-verb אשר ('ashar) generally indicates a decisive progression (Proverbs 4:14, 9:6) or a setting right (Isaiah 1:17). On occasion it's used in the negative (literally: Isaiah 3:12; leading someone "straight astray"), but most often it's positive. So positive even that this verb's secondary meaning is that of being or being made happy (Psalm 41:2, Proverbs 3:18), or even being deemed or called happy (Genesis 30:13, Job 29:11, Psalm 72:16).

The derivatives of this verb are:

- The masculine nouns אשר ('esher) and אשר ('ashar), meaning happiness or blessedness (1 Kings 10:8, Psalm 32:1, Isaiah 30:18). This word most often occurs in the plural construct (that's אשרי or 'happinesses of ...' or 'happinesses to ...', meaning 'happy is ...'), which is not all that odd. Hebrew uses plural to express emphasis, and so, on occasion, does English: 'very, very good times'.
- The masculine noun אשר ('osher), meaning happiness as well, and only used in Genesis 30:13, in the construct באשרי (b'asheray), meaning in my happiness.
- The feminine noun אשור (ashur), meaning a step or a walk; a going (Job 23:11, Psalm 17:4).
- The feminine noun אשר (ashur), also meaning a step or going (Job 31:7, Psalm 17:5, 17:11 only).
- The feminine noun תאשור (te'ashur), denoting a kind of tree, namely the box-tree, which appears to be distinguished by the upward direction of its branches; a happy-tree, or perhaps a straight-up tree (Ezekiel 27:6 only).

אשר

The relative particle אשר (asher), generally meaning who or which, looks like it came straight from the above root, but apparently, that's not so. None of the sources even hints at it, although BDB Theological Dictionary declares its "origin dubious". Our particle occurs in Moabitic with identical meanings but (as HAW Theological Wordbook of the Old Testament notes) it has been found only once in the vast collection of Ugaritic texts that has been unearthed. Since Hebrew and Ugaritic are closely related, this absence of our particle in Ugaritic seems to disarm BDB's objection against one of two plausible theories of its origin:

- This one theory suggests that our word אשר (asher) originated in a word that in Arabic means footstep or mark (which brings it very close to the previous root indeed), then went on to serve as a marker of locality (a place), then acquired the meaning of there and where, and evolved on to become the relative mark we know it as. BDB states that "the chief objection to this explanation is that it would isolate Hebrew from the other Semitic languages, in which pronouns are formed regularly from demonstrative roots".
- The other theory BDB lists involves an unlikely exchange of the letter ל (lamed) of an assumed construction אשל ('sl) for the ר (resh) of our particle אשר (asher). BDB admits that, despite the objection, the previous theory remains most plausible.

The particle אשר (asher) occurs all over the Old Testament (instead of simply submitting a number, HAW Theological Wordbook of the Old Testament excitedly reports that Mandelkern's concordance lists "twenty pages, small print, four columns to each page" of occurrences of אשר (asher).

Our word primarily expresses relation: this *which* that, or he *who* such and such. In some cases it may express result: *so that* if a man could number the dust ... (Genesis 13:16), or purpose: *in order to* find favor (Ruth 2:2), or causality: *because of* their sister (Genesis 34:27), or concession: *although* you made me see trouble (Psalm 71:20).

Our word very often comes with its own preposition, creating even more nuance and meaning:

- With ב (be), meaning in it forms the word באשר (b'sr), which means in which, or in that (Genesis 39:9, Isaiah 56:4).
- With מ (me), meaning from, it forms מאשר (m'sr), which means from that which (Genesis 39:1, Joshua 10:11).
- With the comparative particle כ (ke), meaning like, it forms כאשר (k'sr), which means according as, or simply as (Genesis 34:12, Exodus 10:10, Isaiah 9:2), or it means in so far as or since (Genesis 26:29, Numbers 27:14), or when (Genesis 18:33, 1 Samuel 6:6).

Closely synonymous to the relative particle אש ('asher) is the relative prefix ש (shi). Scholars appear to have concluded that this particle and prefix share no etymological root, but the argumentation surrounding this conclusion is sketchy at best. Whether coincidentally or not, the particle אשר ('asher) and prefix ש (shi) are as alike as the particle על ('al) and the prefix ל (le), and the particle כי (ki) and the prefix כ (ke).

287

ישר

The verb ישר (*yashar*), generally means to be level or straight. It's used in four distinct ways:

- Literally, of a road being straight (1 Samuel 6:12), or smooth (Isaiah 40:3).
- Ethically; of a just or virtuous life style; blameless (Proverbs 11:5), or discerning (Psalm 119:128).
- To be right in the eyes of someone, which means to obtain this person's approval (Judges 14:3).
- Tranquility or harmony: of a soul being at peace (Habakkuk 2:4)

The derivatives of this verb are:

- The adjective ישר (*yashar*), meaning right or upright (Isaiah 26:7, Exodus 15:26).
- The masculine noun ישר (*yosher*), meaning uprightness (Proverbs 2:13, Job 6:25).
- The feminine noun ישרה (*yeshara*), also meaning uprightness (1 Kings 3:6 only).
- The noun מישר (*meshar*), means uprightness, straightness, mostly in an ethical sense (Isaiah 26:7, Proverbs 8:6).
- The noun מישור (*mishor*) means a level place or uprightness mostly in a geographical sense (1 Kings 20:23, Psalm 26:12).

An obvious demonstration of the kinship of these two verbs can be found in the two names Asharelah and Jesharelah, which are applied to the same person.

Associated Biblical names

♂♪	Ahishar	אחישר	♂⊘	Jashar	ישר
♂	Asarel	אשראל	♂	Asharelah-Jesharelah	אשראלה
♂☺	Asher	אשר			ישראלה
♀✱	Asherah	אשרה	♂	Jesher	ישר
		אשירה	○	Jeshurun	ישרון
♂	Asriel	אשריאל	♀○	Sharon	שרון
♂○	Asshur	אשר	○	Telassar	תלאשר
○	Asshurim	אשורם			
○	El-Elohe-Israel	אל אלהי ישראל			
♂○	Israel	ישראל			

Israel meaning

For a meaning of the name Israel, NOBSE Study Bible Name List, BDB Theological Dictionary and Alfred Jones (Dictionary of Old Testament Proper Names) unanimously go with the verb שרה of which the meaning is unsure. Undeterred, NOBSE reads **God Strives**, and BDB proposes **El Persisteth** or **El Persevereth**.

Alfred Jones figures that the mysterious verb שרה might very well mean "to be princely," and assumes that the name Israel consists of a future form of this verb, which hence would mean to become princely. And so Jones interprets the name Israel with **He Will Be Prince With God**.

Here at Abarim Publications, our contention is that the mystery verb שרה doesn't mean struggle at all, but rather reflects a worthiness to govern a nation. At the Jabbok, Jacob became the world's first godly king and his nation was Israel: **God's (Vicarious) Governor**.

SPACE THE TRUE FRONTIER!

21. God is Spirit.

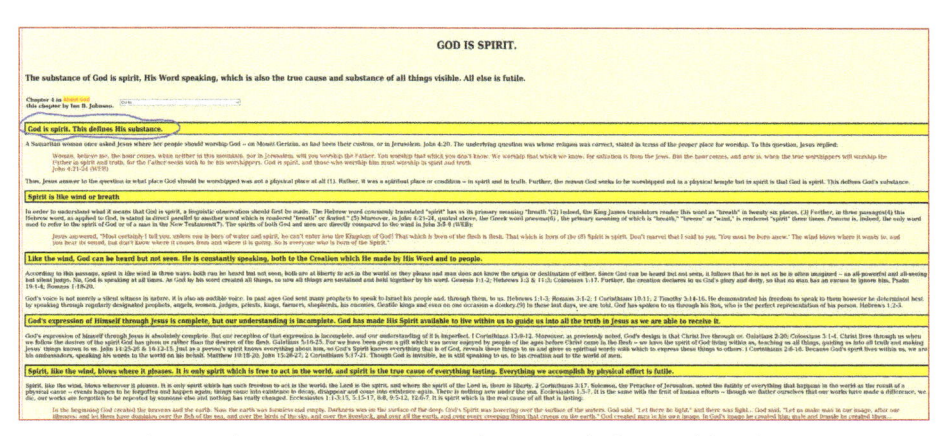

Source: http://christian-oneness.org/about-God/chapter4.html

22. A letter about the Comforter

A friend of mine sent me a letter after I had a discussion about the Spirit of God not being masculine in the New Testament. Below is my reply to him, which I wanted to include.

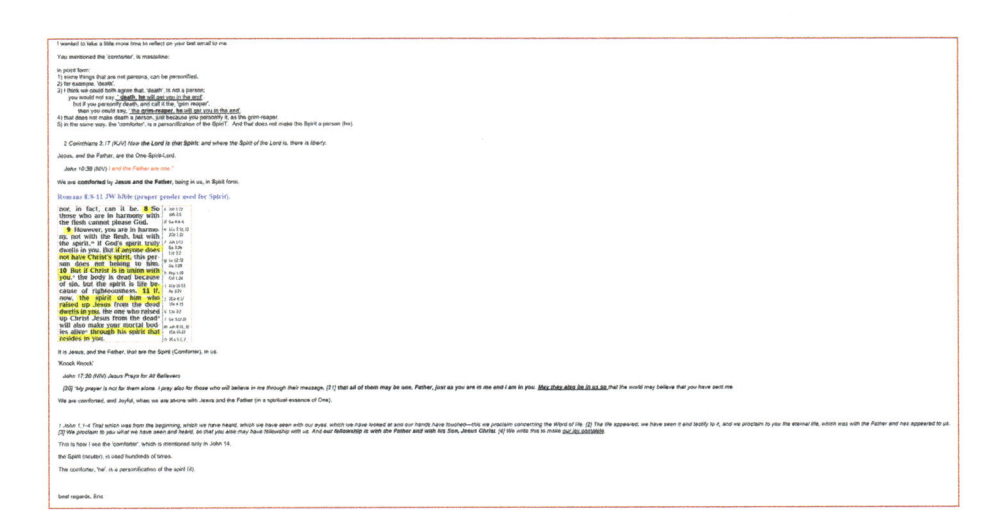

23. Baptised Into Christ

Lesson_17_oes," Are You Really Sure Of Your Eternal Salvation?, accessed November 30, 2017, http://www.netbiblestudy.net/new_page_17.htm.

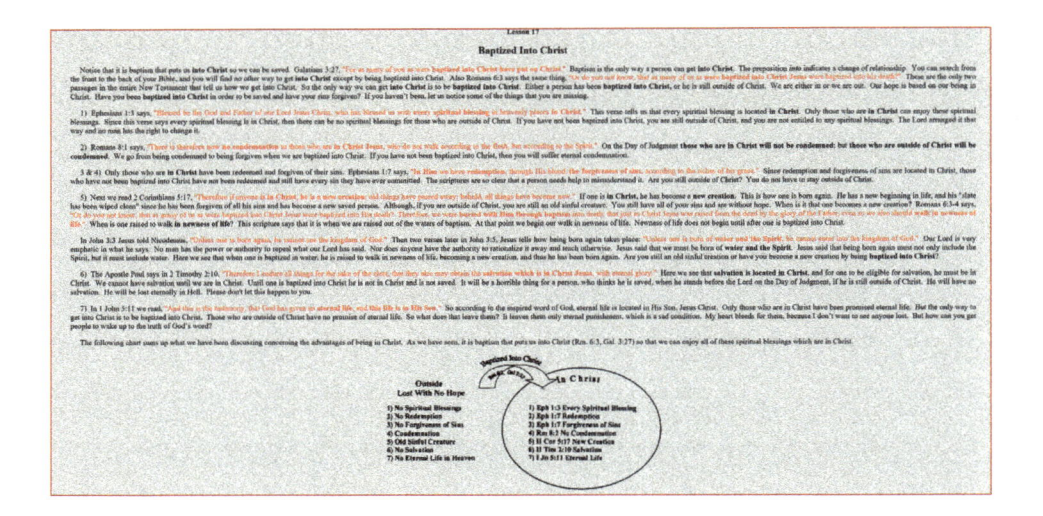

24. Feast days view from an eclectic Internet search that echoes what I am saying and worth considering.

Just who is lying and who is changing things? Who has the whole world deceived? Do you think I do? Do you think some small church does? This is big organization stuff from the prince of the power of the air. How could the whole world be deceived by an insignificant small church or me? Some of these things are in and uphold the word of God, some are not found in the word and involve paganism. You decide which voice is lying. Some people love their paganism and thereby hate, or at least, reject the word of God. (I am sorry some choose the deception, and its comfortable history, even when you should know better.)

*For such men are false apostles, deceitful workmen, masquerading as apostles of Christ. And no wonder, **for***

SPACE THE TRUE FRONTIER!

Satan himself masquerades as an angel of light. It is not surprising, then, if his servants masquerade as servants of righteousness. Their end will be what their actions deserve. (2 Corinthians 11:13–15, NIV)

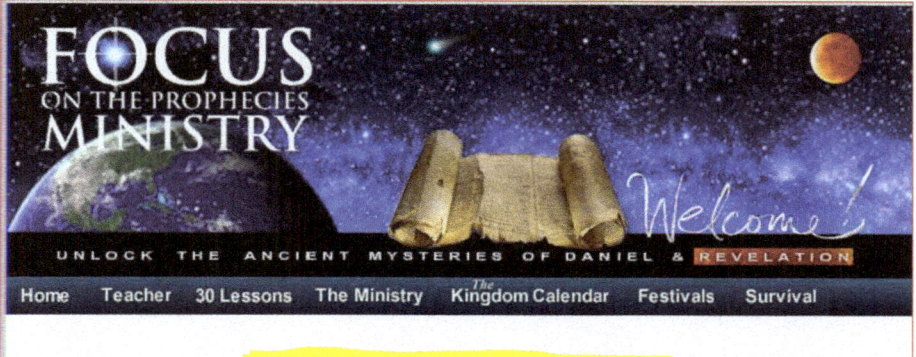

FOCUS ON THE PROPHECIES MINISTRY

UNLOCK THE ANCIENT MYSTERIES OF DANIEL & REVELATION

Home | Teacher | 30 Lessons | The Ministry | The Kingdom Calendar | Festivals | Survival

The Conflict Over God's Appointed "Times & Seasons"
by Kevin R. Swift
June 26, 2012

In the book of Daniel, we discover two very interesting predictions about the final conflict of the ages. It involves God's "times and seasons" or the "set times and the laws" of Almighty God. While Daniel prophesies that our Creator has the ability to "change times and seasons," he also predicts the Man of Sin will, in contrast, attempt to "change the set times and the laws of God. "Change" means "to make the form, nature, content, future course different from what it is; to transform or convert; to substitute; to exchange for something else; to become altered or modified." Clearly, prophecy is asking us to pay attention to these details, and to consider the two opposing sides who are "changing" events, and what this means for the last-day saints.

There are verses in the book of Daniel specifically connected to divine "secrets," which God promises to reveal to the wise and discerning at the appropriate time. Let's look at them. Concerning God, Daniel writes (2:20-22): "Praise be to the name of God for ever and ever; wisdom and power are His. He changes times and seasons; He deposes kings and raises up others. He gives wisdom to the wise and knowledge to the discerning. He reveals deep and hidden things" to those who seek wisdom and knowledge. I'm reminded of a related promise in Daniel 12:9-12, where we are told the wise will understand two important timelines of Bible prophecy when last-day events are about to begin: "Go your way, Daniel, because the words are rolled up and sealed until the time of the end. Many will be purified, made spotless and refined, but the wicked will continue to be wicked. None of the wicked will understand, but those who are wise will understand. "From the time that the daily [worship at the Western Wall/Temple Mount]...is abolished and the abomination that causes desolation is set up [on the Temple Mount], there will be 1,290 days. Blessed is the one who waits for and reaches the end of the 1,335 days." The details have been hidden through the ages, but God is offer wisdom to the last-day saints. It involves His prophetic timelines, along with His plans to change His "times and seasons". It also involves another personality on the end-time stage--an evil actor who opposes truth, and opposes God.

Of this man--the Man of Sin (known also as Antichrist)--Daniel writes (7:25): "He will speak against the Most High and oppress His holy people and try to change the set times and the laws. The holy people will be delivered into his hands for a time, times and half a time," meaning, 3 1/2 years (1,260 days). It's very important that you understand the meaning of the "set times and the laws." The New Living Translation says, "He will try to change their sacred festivals and laws"; and, the Wycliffe Bible says, "He shall think, that he can change the times for the feasts, and the laws." This points to the evil impostor as trying to change God's Festivals, and other Biblical laws found in God's Word.

291

ERIC RASMUSSEN

The Sabbath and the Seven Festivals: Declaring the Past, Foretelling the Future

Long ago God established His "moedim" (meaning, "rehearsals"), referred to as His "set times" and also called "feasts", or seasonal Festivals, which He introduced to the Israelites about 3,400 years ago in Leviticus 23. First, the weekly Sabbath ("These are my appointed festivals, the appointed festivals of the LORD, which you are to proclaim as sacred assemblies. There are six days when you may work, but the seventh day is a day of Sabbath rest, a day of sacred assembly. You are not to do any work; wherever you live, it is a Sabbath to the LORD," verses 2-3). Then follows in the chapter the seven festivals: Passover (Pesach), the Feast of Unleavened Bread (Hag HaMatzah), the Feast of Firstfruits (Bikkurim) and the Festival of Weeks (Shavuot)--these four represent the springtime Festivals. The Feast of Trumpets (Rosh Ha-Shanah), the Day of Atonement (Yom Kippur) and the Feast of Tabernacles (Sukkot) are celebrated in the autumn season. These sanctuary Feasts have commemorated God's historical leading of His people for many centuries, but also, and at the same time, they foretell and predict events about the future. Many centuries after they were established, Yeshua fulfilled the spring Festivals during the week of Passover and at Pentecost. Likewise, He is bound to complete the fall Festivals in His life and redemptive work. Therefore, the autumn Festivals portray the unfolding of live events both in Heaven and on earth, and will be fulfilled by Messiah leading up to and during His Day of the LORD climax of human history. The annual Feasts offer the wise and discerning students of God's Word divine dates of end-time disaster and deliverance--the deep and hidden things that are in process of being revealed.

Paul also connected the "times and seasons" to the Day of the LORD--the coming of Jesus our Messiah and destruction of sinners--when he wrote in 1 Thessalonians 5:1-6: "For you yourselves know perfectly that the Day of the Lord so comes as a thief in the night. For when they say, "Peace and safety!" then sudden destruction comes upon them, as labor pains upon a pregnant woman. And they shall not escape. But you, brethren, are not in darkness, so that this Day should overtake you as a thief. You are all sons of light and sons of the day. We are not of the night nor of darkness. Therefore let us not sleep, as others do, but let us watch and be sober."

Paul indicates that although sinners will not understand or acknowledge the times and seasons, those who are awake need not be in darkness that the Day should overtake God's people as a thief. As sons of light, we are not left in darkness like the world. We are to be sober and wise, knowing the times.

So, this leads to another question. Why will the Man of Sin (Antichrist) attempt to change God's "set times and the laws"? He likely will do so in an attempt to preempt Messiah's final redemption plan that is promised to be completed during the fall Festival season. He will do so by setting himself up in the Temple (on the Temple Mount); then, proclaim himself "GOD" incarnate. Once the Man of Sin sets up the abomination, Messiah's visitation will surely come. Shrouded in darkness, the whole planet will quake at His fury, and the earth will be shaken off it's normal course so that even the "times and seasons" are altered--demonstrating God's supernatural power which Daniel wrote about long ago when he said God "changes times and seasons" (2:20-22).

Paul continued to advise the watching saints on the work of the evil one--the Man of Sin/Lawlessness--when he wrote in 2 Thessalonians 2:1-4: "Concerning the coming of our Lord Jesus Christ and our being gathered to Him, we ask you, brothers and sisters, not to become easily unsettled or alarmed by the teaching allegedly from us—whether by a prophecy or by word of mouth or by letter —asserting that the Day of the Lord has already come. Don't let anyone deceive you in any way, for that Day will not come until the rebellion occurs and the Man of lawlessness is revealed, the man doomed to destruction. He will oppose and will exalt himself over everything that is called God or is worshiped, so that he sets himself up in God's temple, proclaiming himself to be God."

Source: http://www.focusontheprophecies.org/times_and_seasons.htm

25. Christ our Atonement

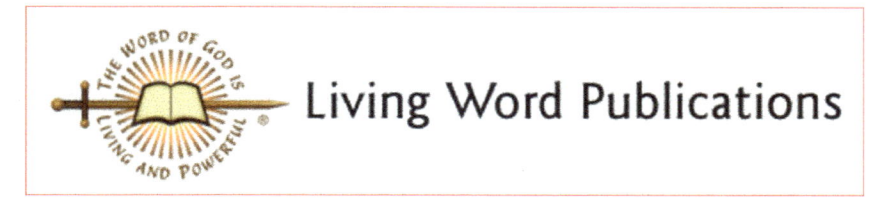

Living Word Publications

Description

Available on CD: Christ Our Atonement, by Gary Hargrave. This message was spoken in North Hills, CA on Sunday, September 15, 2013.

CHRIST OUR ATONEMENT

On the Day of Atonement described in Leviticus 16, the sins and transgressions of the people are literally placed on the goat of removal and taken away. Jesus Christ, our Atonement, not only provides the entire removal of our sin and sin nature but also imparts back to us His divine nature and righteousness. We determine this Day of Atonement concludes with our full appropriation of the total sanctification provided by the Lamb of God.

ABOUT THE AUTHOR

I am married to my wife, Juanita, for just over thirty-eight years. She is my good. We have two boys, both engineers, who are happily married, and I am very proud of them and their mates. We have two wonderful grandchildren that bring us endless joy. And I can't forget our family pets, Calvinator, our ball-fetching, tail-wagging silver Lab eating machine, and our little Mia Capria poodle that I love so much. I want to share my hope with you.